Bogdan-Vasile Matioc

Viscous Fingering in Mathematical Fluid Dynamics via Bifurcation

Bogdan-Vasile Matioc

Viscous Fingering in Mathematical Fluid Dynamics via Bifurcation

A Functional Analytic Approach

Südwestdeutscher Verlag für Hochschulschriften

Impressum/Imprint (nur für Deutschland/ only for Germany)
Bibliografische Information der Deutschen Nationalbibliothek: Die Deutsche Nationalbibliothek
verzeichnet diese Publikation in der Deutschen Nationalbibliografie; detaillierte bibliografische
Daten sind im Internet über http://dnb.d-nb.de abrufbar.

Alle in diesem Buch genannten Marken und Produktnamen unterliegen warenzeichen-, marken-
oder patentrechtlichem Schutz bzw. sind Warenzeichen oder eingetragene Warenzeichen der
jeweiligen Inhaber. Die Wiedergabe von Marken, Produktnamen, Gebrauchsnamen,
Handelsnamen, Warenbezeichnungen u.s.w. in diesem Werk berechtigt auch ohne besondere
Kennzeichnung nicht zu der Annahme, dass solche Namen im Sinne der Warenzeichen- und
Markenschutzgesetzgebung als frei zu betrachten wären und daher von jedermann benutzt
werden dürften.

Verlag: Südwestdeutscher Verlag für Hochschulschriften Aktiengesellschaft & Co. KG
Dudweiler Landstr. 99, 66123 Saarbrücken, Deutschland
Telefon +49 681 37 20 271-1, Telefax +49 681 37 20 271-0
Email: info@svh-verlag.de
Zugl.: Hanover, Gottfried Wilhelm Leibniz University Hanover, Ph D Thesis, 2009.

Herstellung in Deutschland:
Schaltungsdienst Lange o.H.G., Berlin
Books on Demand GmbH, Norderstedt
Reha GmbH, Saarbrücken
Amazon Distribution GmbH, Leipzig
ISBN: 978-3-8381-1259-6

Imprint (only for USA, GB)
Bibliographic information published by the Deutsche Nationalbibliothek: The Deutsche
Nationalbibliothek lists this publication in the Deutsche Nationalbibliografie; detailed
bibliographic data are available in the Internet at http://dnb.d-nb.de.

Any brand names and product names mentioned in this book are subject to trademark, brand
or patent protection and are trademarks or registered trademarks of their respective holders.
The use of brand names, product names, common names, trade names, product descriptions
etc. even without a particular marking in this works is in no way to be construed to mean that
such names may be regarded as unrestricted in respect of trademark and brand protection
legislation and could thus be used by anyone.

Publisher: Südwestdeutscher Verlag für Hochschulschriften Aktiengesellschaft & Co. KG
Dudweiler Landstr. 99, 66123 Saarbrücken, Germany
Phone +49 681 37 20 271-1, Fax +49 681 37 20 271-0
Email: info@svh-verlag.de

Printed in the U.S.A.
Printed in the U.K. by (see last page)
ISBN: 978-3-8381-1259-6

Copyright © 2010 by the author and Südwestdeutscher Verlag für Hochschulschriften
Aktiengesellschaft & Co. KG and licensors
All rights reserved. Saarbrücken 2010

Preface

This work is the thesis I wrote during my four years PhD programme at the Gottfried Wilhelm Leibniz University of Hanover. For this reason I would like to thank first to all my colleagues from the Leibniz University of Hanover for the nice cooperation and the friendly relationship we had during this time.

Above all, I would like to express my gratitude to my advisor, Prof. Dr. Joachim Escher for offering me the chance to study at the Institute of Applied Mathematics. I am grateful for his steady support, encouragement, and for the fruitful discussions and invaluable hints during the last four years. Prof. Escher spent a lot of time checking my manuscripts and our collaboration contributed essentially to my development as a mathematician.

I am also greatly indebted to Prof. Dr. Adrian Constantin for refereeing this thesis and to Prof. Dr. Gheorghe Constantin for guiding me towards Saarbrücken and Hanover, respectively. Especially, I thank my wife Anca for her love, encouragement, and for carefully reading the thesis, and my parents for their support and patience. Finally, I am thankful to the Südwestdeutschen Verlag für Hochschulschriften for the nice and efficient collaboration.

Hanover, December 2009   Bogdan–Vasile Matioc

Contents

1 Introduction   1

2 Modelling aspects and function spaces   7
  2.1 The Navier-Stokes equations 7
  2.2 Banach spaces of Hölder continuous functions 19

I   Well-posedness, stability, and bifurcation results for the flow of a ferrofluid in a rotating Hele–Shaw cell   32

3 The mathematical model   33
  3.1 The physical setting 33
  3.2 The mathematical model 36
  3.3 The transformed problem 42

4 The evolution equation   47
  4.1 The evolution equation 47
  4.2 The well-posedness result 54

5 Stability properties and bifurcation results   58
  5.1 Stability of the trivial equilibrium 58
  5.2 Local bifurcation analysis 64
  5.3 Global bifurcation for analytic operators 70
  5.4 Global bifurcation via Leray-Schauder degree 75

II Well-posedness, analytic dependence, and bifurcation analysis for the Muskat problem　81

6　The mathematical model　82
　6.1　The mathematical model 82
　6.2　Parameterising the boundary 84
　6.3　The transformed system 88

7　The operator equation　92
　7.1　The operator equation 92
　7.2　Proof of Theorem 6.2.2 100
　7.3　Nontrivial steady-states solutions of the
　　　 Muskat problem 114

Bibliography　120

Chapter 1
Introduction

Fingering patterns are complex phenomena which arise at the interface between two fluids or between fluid and air. There are many different situations in which these structures may appear. Saffman and Taylor observed in one of the first works on fingering [58] that whenever the pressure of a less viscous fluid drives forward a more viscous one the interface between them is liable to be unstable and dendrite like structures may occur. Their experiments, in which various fluids were forced into a narrow Hele-Shaw cell, show that single steady shapes with constant velocities of propagation can be produced. It was also pointed out before by Garabedian [36] that there are many possible equilibrium shapes which the finger can have, each of them corresponding to a different velocity of propagation. There is a very rich literature which deals with the Saffman-Taylor fingers (see [62, 65] and the literature therein). The methods used for studying these fluid bubbles rely strongly on the theory of analytic functions since the free boundary of the finger is an analytic curve.

Fingering patterns may also appear also in a horizontal Hele-Shaw cell which rotates with constant angular velocity around its axis. If fluid is injected at the rotation axis of the cell which is opened to air as in [13, 14], or contains another fluid as in [15, 34, 53], then fingering patterns with long, thin radial filaments show off. These experiments led to the conclusion that the instability of the fluid-air interface is driven by the centrifugal forces and may be controlled by the density or viscosity contrast. Similar phenomena appear if one of the fluids is a ferromagnetic and an external magnetic field is applied. This situation is considered in [42, 54, 55] and the authors conclude that the interfacial instabilities which arise can be controlled by the intensity of the magnetic field applied.

The examples presented above can be classified as one- or two-phase moving boundary problems and the fingering patterns are instabilities of some trivial equilibrium of the problem which are due to external actions like gravity, centrifugal forces, an applied pressure gradient etc.

In this thesis we analyse two different situations. The first model we study is a single-phase moving boundary problem describing the evolution of a ferrofluid in a horizontal Hele-Shaw cell. The second one is the dynamic of two wetting phases inside a Hele-Shaw cell, a problem known as the Muskat problem. In both of these cases we deal with volume preserving flows. We prove that the problems are well-posed locally in time for small initial data. As a further common characteristic both models have stationary solutions showing the typical features of fingering.

In the first part of the thesis we consider the evolution of a ferrofluid, surrounded by air at uniform pressure in a horizontal Hele-Shaw cell. The Hele-Shaw cell was invented by the british scientist Henry Selby Hele-Shaw at the end of the nineteenth century. It consists of two transparent, parallel plates separated by a small gap and was used initially for studying laminar flows (see [38]). These cells are also used to simulate and visualise fluid flows in porous media, the growth of crystals, the behaviour of granular materials, to mention just some of their applications. The cell here rotates with constant angular velocity around a rotation axis which is aligned perpendicularly with the centre of the cell. The ferrofluid is exposed to a radial magnetic field generated by the current carried by a wire, which coincides with the rotation axis, and is considered to be of non-Newtonian type. Ferrofluids have a considerable scientific and practical relevance, so that there is an impressive number of physical experiments which deal with this and similar situations (see [42, 52, 54, 64]). Our goal is to determine the motion of the sharp interface between the wetting phase and air. A further unknown is the pressure distribution within the fluids body. Whence, we have to deal with a single-phase moving boundary problem.

The situation when the cell does not rotate and the fluid is Newtonian has been studied by many authors in the last century. The existence of a weak-solution, a global existence and the long-time behaviour of the solution when the initial curve is nearly circular has been discussed in [16]. However, the uniqueness and regularity of that solution is still an open problem. In [3] the authors prove the existence of a unique smooth solution. The solution is constructed as limit of level surfaces of solutions to the

Cahn-Hilliard equation. Local well-posedness results in the Hölder spaces context are found in [23, 30, 31], but in these studies the magnetic field is switched off. The authors construct in [23, 32] centre manifolds consisting of all stationary solutions of the problem which attract at an exponential rate solutions which are initially nearby. If the cell rotates with constant angular velocity, we recently proved in [22] for Newtonian fluids existence and uniqueness of classical solutions. However, the situation is different form that in [16, 31]. Not only that the flow possesses a unique circular stationary solution, but that solution is unstable. Using the surface tension as a bifurcation parameter, the existence of global bifurcation branches consisting of stationary fingering patterns is established. The fingering patterns appear due to the interaction between the surface tension and the centrifugal force. The geometric properties of a special family of stationary fingering patterns has been recently discussed in [50]. Existence and uniqueness of classical solutions and the stability properties of equilibria in strip-like geometries and under various boundary conditions are also found in [24, 25, 26, 27, 28, 29]. Stationary fingering patterns of ferrofluidic substance have been found also in [56], but in an anti-Helmholtz geometry which resembles the situation in [22].

In the second part of the thesis we consider the flow of two fluids in a vertical Hele-Shaw cell with impermeable bottom and laterals. This model is a special case of the Muskat problem when no fluid is injected. The motion of the fluids is determined by the gravity force and the surface tension effects in the interface separating the wetting phases.

The Muskat problem was proposed 1934 by the american engineer Morris Muskat in [55] to describe the flow of two fluids, commonly oil and water, in a porous medium. Though this problem has a weak formulation obtained by Jiang and Chen in [43], the existence of weak solutions is, as far as we know, still open. Local classical solutions of the Muskat problem are first shown to exist by Yi in [66] in a two-dimensional setting and without gravity effects. Using Newton's iteration method, he shows that if the initial speed of the free boundary is positive, then the problem has a unique classical solution on a small time interval. The same author exhibits in the subsequent paper [67] the existence of global classical solutions which are perturbations of a trivial solution that he constructed. Based on complex methods, the authors of [60] show for the stable, forward Muskat problem, in which the higher-viscosity fluid expands into the lower-viscosity fluid, global-in-time existence of solutions for initial data that are small perturbations of a flat interface. For the unstable, backward problem, in which the higher-viscosity fluid contracts, they construct singular solutions that start off with smooth initial data but develop a point of infinite curvature at finite time. A similar result is obtain in [2], where combining energy and complex variable methods the author proves the short-time well-posedness of the Muskat problem in the stable case. It is worth mentioning that the references above do not consider surface tension effects.

Though different in their nature, when solving these two problems we work mainly on the same schema. We shall use, at the beginning, a natural approach for studying moving boundary problems. Namely, we transform the problem on a fixed reference domain. The equation governing the model are more involved now, since nonlinear coefficients are added to them, but we have the great advantage to know the fixed domain we work on. This allows us to introduce solution operators to elliptic boundary value problems associated to our model and to study their regularity properties. These operators are used further on to transform the problems into nonlinear equations on the unit circle.

At this point the problems split, because the Hele-Shaw problem, studied in the first part of the thesis, reduces to an abstract evolution equation which is solved using general results of the theory of parabolic problems as presented in [51]. Furthermore, the principle of linearised stability applies to this situation and permits us to study the stability properties of the trivial circular equilibrium we obtained.

The Muskat model reduces to an operator equation, which is solved by making use of Newton's iteration method. The principle of linearised stability does not apply to this type of equations.

The methods employed for studying the two problems meat again when we consider the set of equilibria of these models. It turns out that the equilibria of both problems are the 2π−periodic solutions of associated nonlinear ODEs. Particularly, the steady-state solutions are all smooth. We use a global bifurcation theorem for analytic operators whose Fréchet derivatives are Fredholm operators of index zero, respectively the global bifurcation theorem due to Rabinowitz, which relies on the Leray-Schauder degree, to find global bifurcation branches consisting of stationary fingering solutions of the problems.

The outline of the thesis is as follows. In the second chapter we illuminate some aspects of the modelling. We deduce the Navier–Stokes equations and show how these equations can be manipulated for flows in porous media or in Hele-Shaw cells to obtain the simpler models we analyse in this thesis. From the momentum of balance equation we obtain, as in [33, 47], a non-Newtonian Darcy law for the gap-averaged velocity. Particularly, for Newtonian fluids we get the well-known (linear) Darcy law. We also deduce the so-called Laplace-Young condition which asserts that the capillary pressure at the interface between two fluids is proportional to the mean curvature of the interface. The function spaces we work with are defined in the second section of this chapter. Of major interest is Theorem 2.2.1 which we first employed in [27] to establish that certain Fourier multiplication operators generate strongly continuous and analytic semigroups in the small Hölder spaces context.

The mathematical model of the problem considered in the first part of this thesis is described in Chapter 3, where we also transform the system of equations we obtained into a problem on a fixed reference domain. In Chapter 4 we re-express the original problem as an abstract evolution equation so that general results from parabolic theory may be applied to obtain the first main result of this work established by Theorems 3.2.3.

We call the rest state when the fluid occupies the unit disc the trivial equilibrium of the problem. As we noticed in Observation 3.2.4 the fluid volume is preserved by the flow. We use this aspect in Chapter 5 when we study the stability properties of this circular steady-state. Theorem 5.1.2 states that the equilibrium is exponentially stable, but only if the current's intensity exceeds a certain critical value. For intensities less than this threshold value the destabilising effect due to angular rotation dominates. If the current's intensity takes exactly this critical value, then the circular steady-state solution attracts at an exponential rate solutions with sufficiently large frequencies.

Finally, we consider in the last section of Chapter 5 the problem of finding all steady-states of the problem. We notice that the equilibria are exactly the $2\pi-$periodic solutions of an associated ODE, and must be therefore smooth. Using the current's intensity as a bifurcation parameter we establish in Theorems 5.2.3 and 5.3.2 the existence of a finite number of global bifurcation branches, which have a local analytic re-parametrisation. If the equilibria on a bifurcation branch remain bounded, but the bifurcation parameter tends to infinity, then the fingering patterns disappear, cf. Corollary 5.3.4. If the surface tension coefficient and the angular velocity satisfy relation (5.14) we show in Theorem 5.4.1 that there exists a unique connected component of the closure of the set of nontrivial solutions of the problem, and this component is unbounded.

In the second part of this thesis we study the flow of two fluids in a vertical Hele-Shaw cell. The mathematical model is presented in Chapter 6. The original problem is transformed in the third section of this chapter on a fixed reference domain by straightening the a priori unknown interface which separates the wetting phases. Chapter 7 is dedicated to the proof of the existence and uniqueness result stated in Theorem 6.2.2. Using interpolation properties of the small Hölder spaces we construct first an approximation of the solution, which enables us to identify the original problem with an operator equation. Theorem 6.2.2 follows then by importing methods form the nonlinear analysis. As a consequence of the implicit function theorem we find out that the solution depends analytically on the initial data.

In the last chapter of this work we determine first that the equilibria of the Muskat problem are the 2π-periodic solutions of a related ODE. The methods used in the last section of Chapter 5 apply very well to the bifurcation problem we obtained. We find an infinite number of global bifurcation branches from the trivial flat solution, cf. Theorem 7.3.2. We have chosen in here the surface tension coefficient to be the bifurcation parameter. As a final result, we show in Corollary 7.3.4 that if the fingering patterns remain bounded and the surface tension goes to infinity along a bifurcation branch, then the fingering patterns flatten out.

Chapter 2
Modelling aspects and function spaces

In this chapter we deduce first the Navier-Stokes equations for Stokesian fluids, which are the starting point to our ongoing work. The balance of momentum equation can be simplified for laminar flows to obtain a non-Newtonian Darcy's law provided the fluid's viscosity satisfies some (very general) conditions. The well-known Laplace-Young condition is also deduced from physical principles.

In the second section of the chapter we introduce Hölder spaces over the unit circle. By identifying the usual Hölder spaces with Besov spaces we prove that Fourier multipliers acting between Hölder spaces of different order and which satisfy some general Marcinkiewicz conditions are continuous.

2.1 The Navier-Stokes equations

Particle paths The mathematical idea of a fluid motion is that it can be described by a point transformation. Let us consider a particle situated at time $t = 0$ at a position ξ, which we call the material coordinates of the particle, and which occupies at time $t > 0$ the position x. The spatial coordinates x

$$x = x(\xi, t) \quad \text{or} \quad x_i = x_i(\xi_1, \xi_2, \xi_3, t) \qquad (2.1)$$

are assumed to depend continuously up to their third derivatives in t and ξ (cf. [7]). Furthermore, it is assumed that equation (2.1) can be inverted to give the initial position of the material coordinates of a particle which is at the position x at time t

$$\xi = \xi(x,t) \quad \text{or} \quad \xi_i = \xi_i(x_1, x_2, x_3, t). \tag{2.2}$$

Denote by J the Jacobian of the transformation (2.1), i.e.

$$J = \frac{\partial(x_1, x_2, x_3)}{\partial(\xi_1, \xi_2, \xi_3)} = \begin{vmatrix} \frac{\partial x_1}{\partial \xi_1} & \frac{\partial x_1}{\partial \xi_2} & \frac{\partial x_1}{\partial \xi_3} \\ \frac{\partial x_2}{\partial \xi_1} & \frac{\partial x_2}{\partial \xi_2} & \frac{\partial x_2}{\partial \xi_3} \\ \frac{\partial x_3}{\partial \xi_1} & \frac{\partial x_3}{\partial \xi_2} & \frac{\partial x_3}{\partial \xi_3} \end{vmatrix}. \tag{2.3}$$

Letting ξ in (2.1) fixed, we can consider (2.1) to be the parametric equation of a curve which goes through ξ at time $t = 0$. These curves are called particle paths. The velocity field at the point $x(\xi, t)$ is defined by the relation

$$v(\xi, t) = \frac{\partial x}{\partial t}(\xi, t).$$

Let us now compute the material derivative $\partial J/\partial t$. Using the Schwarz rule we get

$$\frac{\partial}{\partial t}\left(\frac{\partial x_i}{\partial \xi_j}\right) = \frac{\partial}{\partial \xi_j}\frac{\partial x_i}{\partial t} = \frac{\partial v_i}{\partial \xi_j}. \tag{2.4}$$

On the other hand we have

$$\frac{\partial v_i}{\partial \xi_j} = \frac{\partial v_i}{\partial x_1}\frac{\partial x_1}{\partial \xi_j} + \frac{\partial v_i}{\partial x_2}\frac{\partial x_2}{\partial \xi_j} + \frac{\partial v_i}{\partial x_3}\frac{\partial x_3}{\partial \xi_j} = \frac{\partial v_i}{\partial x_k}\frac{\partial x_k}{\partial \xi_j},$$

if we use the usual summation convention. It is an obvious consequence of the product rule for the derivation and (2.4) that $\partial J/\partial t$ is the sum of three determinants of which the first one is

$$\begin{vmatrix} \frac{\partial v_1}{\partial \xi_1} & \frac{\partial v_1}{\partial \xi_2} & \frac{\partial v_1}{\partial \xi_3} \\ \frac{\partial x_2}{\partial \xi_1} & \frac{\partial x_2}{\partial \xi_2} & \frac{\partial x_2}{\partial \xi_3} \\ \frac{\partial x_3}{\partial \xi_1} & \frac{\partial x_3}{\partial \xi_2} & \frac{\partial x_3}{\partial \xi_3} \end{vmatrix} = \frac{\partial v_1}{\partial x_1}J.$$

Summarising we obtain that

$$\frac{\partial J}{\partial t} = \left(\frac{\partial v_1}{\partial x_1} + \frac{\partial v_2}{\partial x_2} + \frac{\partial v_3}{\partial x_3}\right)J = (\nabla v)J. \tag{2.5}$$

The conservation of mass Consider now an arbitrary volume $V(t)$ of fluid which consists of the same particles at each time t. Let $\rho(x,t)$ be the mass per unit volume of the fluid, assumed to be homogeneous, at position x and time t. For incompressible fluids the density is constant and we denote again by ρ this positive constant. Since the material volume $V(t)$ consists of the same particles, we take it as a principle that the mass does not change, that is

$$\frac{dm}{dt} = \frac{d}{dt}\left(\int_{V(t)} \rho \, dx\right) = 0. \qquad (2.6)$$

Using the diffeomorphism (2.1) we make a change of variables $x = x(\xi, t)$ and rewrite (2.6) in the form

$$\frac{d}{dt}\left(\int_{V(t)} \rho \, dx\right) = \frac{d}{dt}\left(\int_{V_0} \rho J(\xi, t) \, d\xi\right),$$

where $V_0 = V(0)$. We interchange integration with differentiation, and in view of (2.5) we get

$$0 = \int_{V_0} (\nabla v) J \, d\xi.$$

In view of (2.2) we return to the spatial coordinates x and obtain that

$$\int_{V(t)} \nabla v \, dx = 0. \qquad (2.7)$$

Equation (2.7) holds for every volume that could be occupied by the fluid, that is for arbitrary choice of the integration region $V(t)$. We could therefore shrink the region to a point and conclude that the continuous integrand must itself vanish at every point x. We are led to the differential form of the law of conservation of mass

$$\nabla v = 0. \qquad (2.8)$$

The viscosity The forces acting on an element of a continuous medium may be external or body forces, such as gravity, magnetic forces, centrifugal forces or internal (contact forces) which are to be regarded as acting on an element of volume through its bounding surface.

Let n denote the unit outward normal at the surface S of a fluid volume V and t_n the force per unit area exerted there by the material outside S.

The force t_n depends on the position x, the time t and the orientation n. The total internal forces exerted on the volume V with boundary S is

$$\int_S t_n \, ds. \tag{2.9}$$

It holds that $t_n = n \cdot \mathbf{T}$, where $\mathbf{T} = (T_{ij})$ is the so-called stress tensor,

$$\mathbf{T} = -p \cdot \mathrm{id}_{\mathbb{R}^3} + \tau, \tag{2.10}$$

with p the pressure distribution inside the fluids body and τ the extra stress tensor. For a Stokesian or generalised Newtonian fluid the extra stress tensor is given by the relation

$$\tau = 2\mu(\mathrm{tr}(\mathbf{e}^2))\mathbf{e}, \tag{2.11}$$

where $\mathbf{e} := [\nabla v + (\nabla v)^T]/2$ is the deformation (or shear rate) tensor. Letting $\mathbf{e} = (e_{ij})$, we have that $\mathrm{tr}(\mathbf{e}^2) = \sum_{i,j} e_{ij}^2$. We refer here to [8]. Moreover, the mapping

$$\mu : [0, \infty) \to (0, \infty)$$

is called viscosity function. For the remainder of this work we steadily assume that the viscosity function μ of the fluids we consider is real analytic, i.e. $\mu \in C^\omega([0, \infty), (0, \infty))$. If (2.11) is a linear relation the fluid is called Newtonian. Otherwise, if the viscosity function is not constant, the fluid is said to be non-Newtonian. The non-Newtonian fluids are also divided in two classes. The shear thinning fluids, such as ketchup, blood, paint have the property that the viscosity decreases with increasing rate of shear. Otherwise, for shear thickening fluids (suspension of corn flour in water) the viscosity increases with increasing rate of shear.

Balance of momentum The balance of momentum is the first axiom of the classical mechanics, and states that in an inertial frame the rate of change of the momentum of a body is balanced by the force F applied on this body

$$\frac{dP}{dt} = F. \tag{2.12}$$

The momentum of the body is defined by

$$P = \int_{V(t)} \rho v \, dx, \tag{2.13}$$

where $V(t)$ is again a control volume consisting of the same particles. Together with (2.9), we obtain that the whole force acting on the part of fluid in observation is given by

$$F = \int_{V(t)} \mathfrak{f}\, dx + \int_{S(t)} t_n\, ds, \qquad (2.14)$$

where $S(t)$ is the surface enclosing the volume $V(t)$ and the first integral represents the body force acting on $V(t)$. In view of (2.12) and (2.13) we get

$$\frac{d}{dt}\int_{V(t)} \rho v\, dx = \int_{V(t)} \mathfrak{f}\, dx + \int_{S(t)} t_n\, ds,$$

Making a change of variables as in the previous subsection, we obtain in view of (2.5) that

$$\int_V \rho \frac{dv}{dt}\, dx = \int_V \mathfrak{f}\, dx + \int_S t_n\, ds$$

for arbitrary choice of the volume integration V. Here $dv/dt = \partial v/\partial t + v\nabla v$ is the total derivative of v. Plugging in $t_n = n \cdot \mathbf{T}$ we obtain from Stoke's law, by shrinking the integration domain V to a point, the differential form of the balance of momentum:

$$\rho \frac{dv}{dt} + \nabla p = \nabla \tau + \mathfrak{f}. \qquad (2.15)$$

Ferrofluids and the magnetic force A ferrofluid is a liquid which becomes strongly polarised in the presence of a magnetic field. Ferrofluids are colloidal mixtures composed of nanoscale ferromagnetic, or ferrimagnetic, particles ($3 - 15$ nm) coated with a molecular layer of a dispersant (or surfactant) to prevent their agglomeration and suspended in a carrier fluid, usually an organic solvent or water. The particles in a ferrofluid are suspended in Brownian motion, and generally will not settle under normal conditions, even though a slight concentration gradient can become established after long exposure to a force field (gravitational or magnetic). Colloidal ferrofluids must be synthesised since they are not found in Nature.

Although the name may suggest otherwise, ferrofluids do not display ferromagnetism, since they do not retain magnetisation in the absence of an externally applied field. More precisely, in the absence of a magnetic

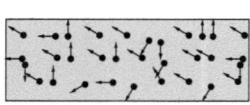 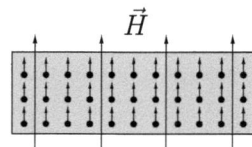

Figure 2.1: Ferroparticles in Brownian motion and in a magnetic field $\vec{H}$.

field, the magnetic moments of the particles are randomly distributed and the fluid has no net magnetisation. When a magnetic field is applied to a ferrofluid, the magnetic moments of the particles orient along the field lines almost instantly (see Figure 2.1). The magnetisation of the ferrofluid responds immediately to the changes in the applied magnetic field and when the applied field is removed, the moments randomise quickly. The body force in ferrohydrodynamics acting on a ferrofluid is due to polarisation, which may be caused by the material magnetisation in the presence of magnetic field gradients.

The intensity of magnetisation M denotes the state of polarisation of magnetised matter. Consider now an isolated small cylindrical volume of magnetically polarised substance with geometric axis $\vec{d}$ aligned with the magnetisation vector $\vec{M}$. The material is subjected to an applied filed $\vec{H}$, of intensity H. The magnetic body force density experienced by the volume element (see [57]) is

$$\mathfrak{f}_m = \mu_0(\vec{M} \cdot \nabla)\vec{H}, \qquad (2.16)$$

where $\mu_0 := 4\pi \cdot 10^{-7}$ is called the permeability of free space.

Ferrofluids have a considerable scientific and practical importance. In many applications, ferrofluid is an active component that contributes towards the enhanced performance of the device. These devices are either mechanical (e.g., seals, bearings and dampers) or electromechanical (e.g., loudspeakers, stepper motors and sensors). Ferrofluids are used to form liquid seals around the spinning drive shafts in hard disks. The rotating shaft is surrounded by magnets. A small amount of ferrofluid, placed in the gap between the magnet and the shaft, will be held in place by its attraction to the magnet. The fluid of magnetic particles forms a barrier which prevents debris from entering the interior of the hard drive.

In medicine, ferrofluids are used as contrast agents for magnetic resonance imaging and can be used for cancer detection. The ferrofluids are in this case composed of iron oxide nanoparticles and called SPION, for "SuperParamagnetic Iron Oxide Nanoparticle". There is also much experimentation with the use of ferrofluids in an experimental cancer treatment called magnetic hyperthermia. It is based on the fact that a ferrofluid placed in an alternative magnetic field releases heat.

The non-Newtonian Darcy law Let us now consider the balance of momentum equation (2.15) in the following context. We presuppose that a volume of fluid is located between the parallel plates of a Hele-Shaw cell, which are either horizontal or vertical, separated by a very small gap b, and having some lateral length scale L such that $\varepsilon := b/L << 1$. We further assume that there exists a potential function u such that

$$\nabla u = \nabla p - \mathfrak{f}, \tag{2.17}$$

where $\mathfrak{f}$ denotes the resultant body force to which the liquid is exposed to. This is definitely the case when a incompressible fluid blob, which at time t occupies the domain $V(t)$, is located in a vertical Hele-Shaw cell and the only external force acting on it is the gravitational force $\mathfrak{f}_g = (0, -g\rho, 0)$. Here g is the gravitational constant and ρ the fluid's density. In this case

$$u = p + g\rho y, \tag{2.18}$$

with y one of the lateral coordinates and z the gap coordinate (see Figure 2.2).

The second situation we may consider in this context is that of a ferrofluid situated in a Hele-Shaw cell which rotates with angular velocity in the horizontal plane. The gravity effects are neglected and the centrifugal and magnetic forces are the only external forces acting on the bulk fluid. The situation is described in the first part of this work and the potential u is explicitly given explicitly by relation (3.1).

Under these assumptions, the balance of momentum equation (2.15) writes in both situations

$$\rho \frac{dv}{dt} + \nabla u = \nabla \tau. \tag{2.19}$$

In the thin gap between the plates the Reynolds number is small and inertial terms can be neglected (see [33, 47]) to obtain that

$$\nabla u = \nabla \tau = 2\nabla(\mu(\mathrm{tr}(\mathbf{e}^2))\mathbf{e}), \tag{2.20}$$

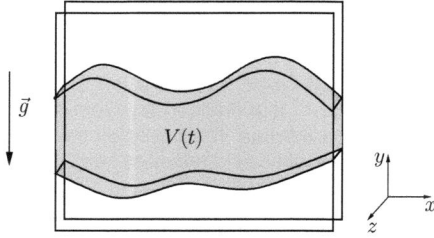

Figure 2.2: Fluid blob in a vertical Hele-Shaw cell

where we use the notations previously established. The Navier-Stokes equations are further simplified by making use of the large aspect of the cell. Let $x = Lx', y = Ly', z = bz'$ and $\varepsilon = b/L << 1$. The velocity is scaled correspondingly, i.e. if $v = (v_1, v_2, v_3)$, we have that

$$v_i(x, y, z) = L v'_i(x', y', z'), \ i = 1, 2,$$

$$v_3(x, y, z) = b v'_3(x', y', z').$$

It follows that

$$\mathbf{e}^2 = \frac{\varepsilon^{-2}}{2}(v'_1, v'_2)_{z'} \cdot (v'_1, v'_2)_{z'} + O(1, \varepsilon).$$

Retaining only lowest order terms in (2.20) yields the reduced equations

$$\nabla_2 u = \partial_z \left[\mu \left(\frac{|\overline{v}_z|^2}{2} \right) \overline{v}_z \right], \quad \partial_z u = 0, \qquad (2.21)$$

where $\overline{v} = (v_1, v_2)$ and $\nabla_2 = (\partial_x, \partial_y)$. We continue without further approximation. As u depends only upon x and y, equation (2.21) integrates to

$$z \nabla_2 u = \mu \left(\frac{|\overline{v}_z|^2}{2} \right) \overline{v}_z, \qquad (2.22)$$

if we consider symmetric velocity profiles about $z = 0$. As in the usual Darcy's law, we wish to find $\overline{v}$ as a function of $\nabla_2 u$. Squaring the equation

we obtain that
$$\frac{z^2|\nabla_2 u|^2}{2} = \mu^2\left(\frac{|\bar{v}_z|^2}{2}\right)\frac{|\bar{v}_z|^2}{2} =: h\left(\frac{|\bar{v}_z|^2}{2}\right), \tag{2.23}$$
where
$$h : [0, \infty) \to [0, \infty), \qquad h(r) := r\mu^2(r) \quad \text{for } r \geq 0. \tag{2.24}$$
Our goal is to invert the relation (2.23). Therefore, we assume that the viscosity μ satisfies the conditions
$$0 < \mu(r) \quad \text{for all } r \geq 0,$$
$$0 < \mu(r) + 2r\mu'(r) \quad \text{for all } r \geq 0, \tag{2.25}$$
$$r\mu^2(r) \to_{r \to \infty} \infty.$$

We point out here that the first two conditions on the viscosity function were also found in [10], where the full Navier-Stokes problem but on a fixed domain is studied. These assumptions ensure the invertibility of the mapping h. They are very general and apply to many fluids (see the examples below). We formulate also the stronger version of relations (2.25)
$$m_\mu \leq \mu \leq M_\mu \quad \text{and} \quad m_\mu \leq \mu(r) + 2r\mu'(r) \leq M_\mu \tag{2.26}$$
for $r \geq 0$, where m_μ and M_μ are positive constants.

That the conditions (2.25) and (2.26) are very general can be seen from the large number of examples we give below. For the Oldroyd-B fluids, e.g. blood, the viscosity is given by
$$\mu(r) = \nu_\infty + (\nu_0 - \nu_\infty)\frac{1 + \ln(1 + \lambda r)}{1 + \lambda r}, \quad r \geq 0,$$
where $\lambda > 0$ is a material constant and $\nu_0 > \nu_\infty > 0$. The conditions (2.26) hold if $(e^2 + 1)\nu_\infty > \nu_0$. Also, various power law fluids can be considered. Let
$$\mu(r) = \nu_\infty + \nu_0(1 + r^2)^{s/4}, \quad \text{or} \quad \mu(r) = \nu_\infty + \nu_0(1 + r)^{s/2},$$
for all $r \geq 0$, where ν_0 and ν_∞ are positive and $s \leq 0$. In this case (2.26) hold if $0 \geq s \geq -1$. A further interesting example is the Bingham model with dynamic viscosity
$$\mu(r) := \mu_\infty + \frac{\beta \tau_0}{1 + \beta r},$$

where μ_∞, τ_0 are positive constants and $\beta \geq 0$ (see [63]). Relation (2.26) is satisfied iff $\beta\tau_0 < 8\mu_\infty$. For the Johnson-Segalman-Oldroyd model with relaxation times λ_k and k-th mode viscosities η_k, $k = 1, ..., N$, the viscosity function is given by

$$\mu(r) = \alpha_0 + \sum_{k=1}^{N} \frac{\alpha_k}{1 + \beta_k^2 \cdot r},$$

where the positive constants occurring in the relation are assumed to satisfy

$$\alpha_0 = \mu_s/\mu_0, \quad \alpha_k = \eta_k/\mu_0, \quad \beta_k = \lambda_k/\lambda_1, \text{ and } \mu_0 = \mu_s + \sum_{k=1}^{N} \eta_k.$$

We refer to [33] for more details. In this case the relation (2.26) hold for any choice of the parameters. Notice that these fluids are all shear thinning.

We give now an example of a shear thickening fluid which fits into our context. If

$$\mu(r) = \mu_0 \frac{\gamma r + r_0}{r + r_0}, \quad \forall r \geq 0,$$

with $r_0 > 0$, $\gamma \geq 1$, and $\mu_0 > 0$, then (2.26) hold for any choice of the parameters r_0, μ_0, and γ.

We have seen that conditions (2.26) are not to restrictive. In addition to the examples mentioned above, we include also the following cases

$$\mu(r) = \nu_0(1 + r)^{s/2} \quad \text{and} \quad \mu(r) = \nu_\infty + \nu_0 r^{s/2},$$

where ν_0 and ν_∞ are positive constants. These are both power-law fluids. For the first example the parameter s must belongs to $(-1, \infty)$ to ensure that (2.25) hold true. The second example was introduced in the mathematical literature by Ladyzhenskaya in [48]. Here, (2.25) are satisfied iff $s \in \{0\} \cup [2, \infty)$. Notice that relations (2.26) are not verified by these fluids if $s > 0$.

Back to equation (2.23). We invert it and get $|\bar{v}_z|^2/2 = h^{-1}(z^2|\nabla_2 u|^2/2)$, and substituting into (2.22) yields

$$\bar{v}_z = \frac{z\nabla_2 u}{\widetilde{\mu}\left(\frac{z^2|\nabla_2 u|^2}{2}\right)} \quad \text{or} \quad \bar{v} = \int_{-b}^{z} \frac{z'\nabla_2 u}{\widetilde{\mu}\left(\frac{z'^2|\nabla_2 u|^2}{2}\right)} dz', \qquad (2.27)$$

where $\widetilde{\mu} := \mu \circ h^{-1}$. We have assumed no-slip on the lateral plates, i.e. $v = 0$ on $z = \pm b$. The gap-averaged velocity is defined by the relation

$$\vec{v}(x,y) = \frac{1}{2b} \int_{-b}^{b} \overline{v}(x,y,z)\,dz. \tag{2.28}$$

Gap averaging equation (2.27) we compute, using Fubini's theorem, that

$$\vec{v} = \frac{1}{2b} \int_{-b}^{b} \int_{-b}^{z} \frac{z' \nabla_2 u}{\widetilde{\mu}\left(\frac{z'^2 |\nabla_2 u|^2}{2}\right)}\,dz'\,dz = \frac{1}{2b} \int_{-b}^{b} \int_{z'}^{b} \frac{z' \nabla_2 u}{\widetilde{\mu}\left(\frac{z'^2 |\nabla_2 u|^2}{2}\right)}\,dz\,dz'$$

$$= \frac{1}{2b} \int_{-b}^{b} \frac{z'(b-z') \nabla_2 u}{\widetilde{\mu}\left(\frac{z'^2 |\nabla_2 u|^2}{2}\right)}\,dz' = -\left(\frac{1}{2b} \int_{-b}^{b} \frac{z'^2}{\widetilde{\mu}\left(\frac{z'^2 |\nabla_2 u|^2}{2}\right)}\,dz'\right) \nabla_2 u.$$

The substitution $z' = \sqrt{2}s$ yields a non-Newtonian Darcy's law for the gap averaged velocity in a two-dimensional setting

$$\vec{v} = -\frac{\nabla u}{\overline{\mu}(|\nabla u|^2)}, \tag{2.29}$$

where letting $\overline{b} := b/\sqrt{2}$ and $\overline{c} = \sqrt{2}/b$, we defined the effective viscosity $\overline{\mu}$ by the relation

$$\frac{1}{\overline{\mu}(r)} = \overline{c} \int_{-\overline{b}}^{\overline{b}} \frac{s^2}{\widetilde{\mu}(s^2 r)}\,ds. \tag{2.30}$$

Since μ is a real analytic function, it is obvious that the effective viscosity $\overline{\mu}$ is smooth. Notice also, that if the viscosity μ is constant, then we obtain the Newtonian version of Darcy's law

$$\vec{v} = -\frac{\overline{k}}{\mu} \nabla u, \tag{2.31}$$

where $\overline{k} = b^2/3$. Gap averaging the conservation of mass equation (2.8) we also get that

$$\nabla \vec{v} = 0. \tag{2.32}$$

Free surfaces and the Laplace-Young condition At a free surface separating two fluids from each other the phenomenon of surface or capillary tension is present. The surface tension is very important in technical problems since it can affect the stability properties of the equilibria (see [26, 28]), help to prove the existence of nontrivial equilibria for complex free boundary problems (see [18, 68]), or even determine the well-posedness of moving boundary problems (for example the Muskat problem considered in the second part of this work). The surface tension is due to the fact that molecules on the free surface, or an interface between two fluids are in a different environment than those molecules within a fluid. They do not have other like molecules on all sides of them, and consequently they cohere more strongly to those directly associated with them on the surface.

The capillary force on a line element of this free surface is

$$\Delta F = \vec{\gamma} \Delta s,$$

where $\vec{\gamma}$ is the stress vector of the surface tension, defined by

$$\vec{\gamma} = \lim_{\Delta s \to 0} \frac{\Delta F}{\Delta s} = \frac{dF}{ds}.$$

If the surface is at rest, the stress vector acting on a line contained in the interface is

$$\vec{\gamma} = \gamma N,$$

where N is the normal to the line which is parallel to the surface, i.e. the capillary force is parallel to the surface but perpendicular to the line. We call γ surface tension coefficient.

Let S be a portion of a surface separating two fluids, N the outward normal at ∂S, n the outward normal at the surface S, and $t = N \times n$ a unit vector tangent at ∂S (see Figure 2.3). At equilibrium, the force due to surface tension acting on ∂S is compensated by the internal forces acting on S

$$\int_{\partial S} \gamma N \, ds = \int_S (t_{n-} + t_{n+}) \, dS,$$

cf. (2.9), where we have denoted by t_{n-} and t_{n+} the internal for exerted on S by each of the fluids, respectively. In view of (2.10) we get

$$\int_{\partial S} \gamma N \, ds = \int_S (p_+ - p_-) n \, dS.$$

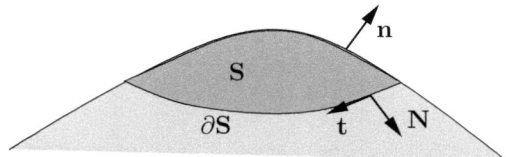

Figure 2.3: A local geometry

Using Stokes' theorem as in [11, Exercise 6.15], we get, in view of $\nabla n = -\kappa_S$, that

$$\int_S (p_+ - p_-) n \, dS = -\gamma \int_{\partial S} n \times t \, ds = -\gamma \int_{\partial S} n \times d\mathbf{s}$$

$$= \gamma \int_S (n \times \nabla) \times n \, dS = \gamma \int_S \left(\nabla(n \cdot n) - n(\nabla \cdot n) \right) dS$$

$$= -\gamma \int_S n(\nabla \cdot n) \, dS = \gamma \int_S \kappa_S n \, dS,$$

where κ_S stands for the mean curvature of Γ_S. By shrinking S to a point, we obtain, at each point of the surface separating two fluids, the Laplace-Young condition

$$p_+ - p_- = \gamma \kappa_S, \tag{2.33}$$

which asserts that the pressure jump across the interface is compensated by the mean curvature of the surface.

2.2 Banach spaces of Hölder continuous functions

We denote by $\mathbb{S}$ the unit circle $\mathbb{S} := \{x \in \mathbb{R}^2 : |x| = 1\}$. Functions defined on $\mathbb{S}$ are naturally identified with 2π−periodic functions on $\mathbb{R}$. Where there is no danger of confusion, we shall denote also by x the real variable.

The space $C^m(\mathbb{S})$, $m \in \mathbb{N}$, consists of functions whose m-th order derivatives are continuous, and is a Banach space with the norm

$$\|f\|_{C^m(\mathbb{S})} := \sum_{k=0}^{m} \max_{\mathbb{S}} |f^{(k)}|.$$

Given $m \in \mathbb{N}$ and $\beta \in (0,1)$, the space $C^{m+\beta}(\mathbb{S})$ consists of the 2π–periodic and real-valued functions on $\mathbb{R}$ which are m–times continuously differentiable and additionally

$$[f]_{m,\beta} := \sup_{\substack{x,x' \in \mathbb{R} \\ x \neq x'}} \frac{|f^{(m)}(x) - f^{(m)}(x')|}{|x-x'|^\beta} < \infty.$$

Endowed with the norm

$$\|f\|_{C^{m+\beta}(\mathbb{S})} := \|f\|_{C^m(\mathbb{S})} + [f]_{m,\beta} \quad \text{for} \quad f \in C^{m+\beta}(\mathbb{S}),$$

the Hölder space $(C^{m+\beta}(\mathbb{S}), \|\cdot\|_{C^{m+\beta}(\mathbb{S})})$ is a Banach space.

By using dyadic decomposition, it is shown in [59, Theorem 3.5.4 (i)] that, if $r > 0$ is not an integer, then $C^r(\mathbb{S}, \mathbb{C}) := B^r_{\infty,\infty}(\mathbb{S})$. Given $r \geq 0$, $C^r(\mathbb{S}, \mathbb{C}) := C^r(\mathbb{S}) + iC^r(\mathbb{S})$ is the complexification of $C^s(\mathbb{S})$. The Besov spaces $B^s_{\infty,\infty}(\mathbb{S})$ are defined as follows. We use the standard notation $\mathcal{D}(\mathbb{S}, \mathbb{C})$ when we refer to the space consisting of complex-valued smooth functions on $\mathbb{S}$, i.e. $\mathcal{D}(\mathbb{S}, \mathbb{C}) = C^\infty(\mathbb{S}, \mathbb{C})$. Given $n \in \mathbb{N}$, the function

$$\|\varphi\|_{C^n(\mathbb{S},\mathbb{C})} := \max_{k \leq n} \max_{\mathbb{S}} |\varphi^{(k)}|$$

is norm on $\mathcal{D}(\mathbb{S}, \mathbb{C})$, and endowed with the metric

$$d(\varphi, \psi) := \sum_{n=0}^{\infty} \frac{1}{2^n} \frac{\|\varphi - \psi\|_{C^n(\mathbb{S},\mathbb{C})}}{1 + \|\varphi - \psi\|_{C^n(\mathbb{S},\mathbb{C})}}, \quad \varphi, \psi \in \mathcal{D}(\mathbb{S}, \mathbb{C}),$$

$\mathcal{D}(\mathbb{S}, C)$ is a Fréchet space. Let $\mathcal{D}'(\mathbb{S}, \mathbb{C})$ denote the topological dual of $\mathcal{D}(\mathbb{S}, \mathbb{C})$. Obviously, $\mathcal{D}(\mathbb{S}, \mathbb{C}) \subset \mathcal{D}'(\mathbb{S}, \mathbb{C})$. Given $f \in \mathcal{D}'(\mathbb{S}, \mathbb{C})$, the complex numbers

$$\widehat{f}(k) = f(e^{-ikx}), \, k \in \mathbb{Z}$$

are the Fourier coefficients of the linear functional f. The following facts are well-known, cf. [59]. Any function $\varphi \in \mathcal{D}(\mathbb{S}, \mathbb{C})$ can be represented by its Fourier series

$$\varphi = \sum_{k \in \mathbb{Z}} \widehat{\varphi}(k) e^{ikx},$$

which converges in $\mathcal{D}(\mathbb{S}, \mathbb{C})$ and

$$|\widehat{\varphi}(k)| \leq c_m(1+|k|)^{-m}, \quad k \in \mathbb{Z} \tag{2.34}$$

for all $m \in \mathbb{N}$. Conversely, if $\varphi = \sum_{k \in \mathbb{Z}} a_k e^{ikx}$, and $(a_k)_{k \in \mathbb{Z}}$ satisfies (2.34), then $\varphi \in \mathcal{D}(\mathbb{S}, \mathbb{C})$. Analogously, any distribution $f \in \mathcal{D}'(\mathbb{S}, \mathbb{C})$ can be represented as

$$f = \sum_{k \in \mathbb{Z}} \widehat{f}(k) e^{ikx}.$$

The series converges in $\mathcal{D}'(\mathbb{S}, \mathbb{C})$ and

$$|\widehat{f}(k)| \leq c_m(1+|k|)^{m}, \quad k \in \mathbb{Z} \tag{2.35}$$

for some $m \in \mathbb{N}$. Vice versa, the series $\sum_{k \in \mathbb{Z}} a_k e^{ikx}$ converges in $\mathcal{D}'(\mathbb{S}, \mathbb{C})$, provided $(a_k)_{k \in \mathbb{Z}}$ satisfies (2.35) for some $m \in \mathbb{N}$.

Let $(\phi_j)_{j \geq 0}$ be a sequence in the Schwartz space $\mathcal{S}(\mathbb{R})$ with the following properties:

(i) $\operatorname{supp} \phi_0 \subset [-2, 2]$, $\quad \operatorname{supp} \phi_j \subset \{x : 2^{j-1} \leq |x| \leq 2^{j+1}\}, j \geq 1$;

(ii) $\displaystyle\sum_{j \in \mathbb{N}} \phi_j = 1$, on $\mathbb{R}$;

(iii) $\forall k \in \mathbb{N} \exists c_k > 0 : \quad 2^{kj} \|\phi_j^{(k)}\|_0 \leq c_k, \forall j \in \mathbb{N}$.

For $s > 0$ we define

$$B_{\infty,\infty}^s(\mathbb{S}) := \left\{ f = \sum_{k \in \mathbb{Z}} \widehat{f}(k) e^{ikx} \in \mathcal{D}'(\mathbb{S}, \mathbb{C}) : \right.$$

$$\left. \|f\|_{B_{\infty,\infty}^r(\mathbb{S})}^{(\phi_j)} := \sup_{j \in \mathbb{N}} 2^{sj} \left\| \sum_{k \in \mathbb{Z}} \phi_j(k) \widehat{f}(k) e^{ikx} \right\|_{C(\mathbb{S}, \mathbb{C})} < \infty \right\},$$

the so-called Besov space. It can be shown, cf. [59] that $B_{\infty,\infty}^s(\mathbb{S})$ is a Banach space. Moreover, the norm $\|\cdot\|_{B_{\infty,\infty}^s(\mathbb{S})}^{(\phi_j)}$ depends on the sequence (ϕ_j) in the following sense. If $(\psi_j)_{j \geq 0} \subset \mathcal{S}(\mathbb{R})$ satisfies the conditions $(i) - (iii)$ mentioned above, then the norm $\|\cdot\|_{B_{\infty,\infty}^s(\mathbb{S})}^{(\psi_j)}$ is equivalent to $\|\cdot\|_{B_{\infty,\infty}^s(\mathbb{S})}^{(\phi_j)}$. This description of the Hölder spaces using dyadic decomposition is very useful because it simplifies things when one tries to characterise Fourier multiplication operators between Hölder spaces. The main result of this section is the following theorem, found first in [6] and used to characterise operator valued Fourier multiplication operators on periodic Besov spaces, but in the special case when the Besov spaces have the same order. It was generalised in [26] for Fourier multiplication operators between complex valued periodic Besov spaces of different order and used to prove that the linearisation of some nonlinear and nonlocal operator generates a strongly continuous and analytic semigroup.

Before stating the theorem we recall the definition of Fourier multiplication operators. Let $(M_k)_{k\in\mathbb{Z}}$ be a sequence of complex numbers, which we call the symbol. The Fourier multiplication operator associated to the symbol $(M_k)_{k\in\mathbb{Z}}$ is the operator

$$\sum_{k\in\mathbb{Z}} \widehat{f}(k)e^{ikx} \xmapsto{T} \sum_{k\in\mathbb{Z}} M_k \widehat{f}(k)e^{ikx}.$$

It is quite elementary to show that if $(M_k)_k$ is a bounded sequence, then $T \in \mathcal{L}(L^2(\mathbb{S},\mathbb{C}))$. Here, $L^2(\mathbb{S},\mathbb{C})$ stands for the space of square integrable functions on $\mathbb{S}$. The situation when T acts between spaces of continuous functions is more subtle, as the next theorem shows.

Theorem 2.2.1 *Let r, s be two positive real numbers and let $(M_k)_{k\in\mathbb{Z}} \subset \mathbb{C}$ be a sequence satisfying the following conditions*

$(i)\quad s_1 := \sup_{k\in\mathbb{Z}\setminus\{0\}} |k|^{r-s}|M_k| < \infty,$

$(ii)\quad s_2 := \sup_{k\in\mathbb{Z}\setminus\{0\}} |k|^{r-s+1}|M_{k+1} - M_k| < \infty,$

$(iii)\quad s_3 := \sup_{k\in\mathbb{Z}\setminus\{0\}} |k|^{r-s+2}|M_{k+2} - 2M_{k+1} + M_k| < \infty.$

Then the mapping

$$\sum_{k\in\mathbb{Z}} \widehat{f}(k)e^{ikx} \xmapsto{T} \sum_{k\in\mathbb{Z}} M_k \widehat{f}(k)e^{ikx},$$

belongs to $\mathcal{L}(B^s_{\infty,\infty}(\mathbb{S}), B^r_{\infty,\infty}(\mathbb{S}))$.

The relations $(i) - (iii)$ are called general Marcinkiewicz conditions. Before starting the proof of Theorem 2.2.1, we give first an auxiliary lemma. Though elementary, this lemma plays an important role both in the proof of Theorem 2.2.1 and of the well-posedness result stated in Theorem 6.2.2 in the second part of this work. Let $\mathcal{F}$ and $\mathcal{F}^{-1}$ denote the Fourier transform and the inverse Fourier transform operator, respectively, i.e.

$$\mathcal{F}f(\xi) = (2\pi)^{-1/2} \int_{\mathbb{R}} f(x) e^{-ix\xi} \, dx,$$

and $\mathcal{F}^{-1}f(\xi) = \mathcal{F}f(-\xi)$ for $f \in L^1(\mathbb{R}, \mathbb{C})$ and $\xi \in \mathbb{R}$. As usual, $L^1(\mathbb{R}, \mathbb{C})$ stands for the space of the Lebesgue integrable functions on $\mathbb{R}$. It is well known (cf. [5, Corollary 9.13]) that if $f \in C(\mathbb{R}, \mathbb{C})$ is compactly supported and $\mathcal{F}^{-1}f \in L^1(\mathbb{R}, \mathbb{C})$, then $f = \mathcal{F}\mathcal{F}^{-1}f$.

Lemma 2.2.2 *Let* $\mathrm{m} \in C(\mathbb{R}, \mathbb{C})$ *be a function with compact support with the property that* $\mathcal{F}^{-1}\mathrm{m} \in L^1(\mathbb{R}, \mathbb{C})$. *Given a trigonometric polynomial*

$$f = \sum_{k \in \mathbb{Z}} \widehat{f}(k) e^{ikx}$$

we have that

$$\left\| \sum_{k \in \mathbb{Z}} \mathrm{m}(k) \widehat{f}(k) e^{ikx} \right\|_{C(\mathbb{S}, \mathbb{C})} \leq (2\pi)^{-1/2} \|\mathcal{F}^{-1}\mathrm{m}\|_{L^1} \|f\|_{C(\mathbb{S}, \mathbb{C})}. \tag{2.36}$$

Proof Being a trigonometric polynomial, the Fourier coefficients of f satisfy $\widehat{f}(k) = 0$ provided k is large enough. By interchanging integration and summation we get for $x \in \mathbb{R}$

$$\sum_{k \in \mathbb{Z}} \mathrm{m}(k) \widehat{f}(k) e^{ikx} = \sum_{k \in \mathbb{Z}} \mathcal{F}(\mathcal{F}^{-1}\mathrm{m})(k) \widehat{f}(k) e^{ikx}$$

$$= (2\pi)^{-1/2} \sum_{k \in \mathbb{Z}} \int_{\mathbb{R}} \mathcal{F}^{-1}\mathrm{m}(y) e^{ik(x-y)} \, dy \, \widehat{f}(k)$$

$$= (2\pi)^{-1/2} \int_{\mathbb{R}} \mathcal{F}^{-1}\mathrm{m}(y) \sum_{k \in \mathbb{Z}} \widehat{f}(k) e^{ik(x-y)} \, dy$$

$$= (2\pi)^{-1/2} \int_{\mathbb{R}} \mathcal{F}^{-1}\mathrm{m}(y) f(x-y) \, dy$$

$$= (2\pi)^{-1/2} \mathcal{F}^{-1}\mathrm{m} * f(x),$$

where $\mathcal{F}^{-1}\mathrm{m} * f$ stands for the convolution of $\mathcal{F}^{-1}\mathrm{m}$ and f. The statement follows now directly from Young's inequality. □

The inequality (2.36) is an important tool in proving Theorem 2.2.1. It reduces the proof of the theorem to seeking uniform estimates for the $L^1(\mathbb{R}, \mathbb{C})$–norm of countably many continuous functions with compact support.

Proof (Proof of Theorem 2.2.1) Let $r, s \in (0, \infty)$ and $(M_k)_{k \in \mathbb{Z}} \subset \mathbb{C}$ be a sequence for which the general Marcinkiewicz conditions $(i) - (iii)$ hold. From the definition of the Besov norm it suffices to find a constant $C > 0$ such that

$$2^{rj} \left\| \sum_{k \in \mathbb{Z}} \phi_j(k) M_k \widehat{f}(k) e^{ikx} \right\|_{C(\mathbb{S},\mathbb{C})} \leq C 2^{sj} \left\| \sum_{k \in \mathbb{Z}} \phi_j(k) \widehat{f}(k) e^{ikx} \right\|_{C(\mathbb{S},\mathbb{C})}$$

for all $j \in \mathbb{N}$. We are thus looking for a positive constant C, independent of f, such that

$$\left\| \sum_{k \in \mathbb{Z}} \phi_j(k) \left(2^{(r-s)j} M_k\right) \widehat{f}(k) e^{ikx} \right\|_{C(\mathbb{S},\mathbb{C})} \leq C \left\| \sum_{k \in \mathbb{Z}} \phi_j(k) \widehat{f}(k) e^{ikx} \right\|_{C(\mathbb{S},\mathbb{C})}$$

for all $j \in \mathbb{N}$ and $f \in \mathcal{D}'(\mathbb{S}, \mathbb{C})$.

Given $j \in \mathbb{N}, j \geq 1$, we define the piecewise affine function $\mathrm{m}_j : \mathbb{R} \to \mathbb{C}$ by $\mathrm{m}_j = 0$ on $[|x| \leq 2^{j-2}] \cup [|x| \geq 2^{j+2}]$, $\mathrm{m}_j(k) = 2^{(r-s)j} M_k$ for $2^{j-1} \leq |k| \leq 2^{j+1}$ and additionally m_j is affine on each interval $[k, k+1], k \in \mathbb{Z}$ (see Figure 2.2 where m_j is represented for $(M_k)_{k \in \mathbb{Z}} \subset \mathbb{R}$).

For $j = 0$, the mapping m_0 is defined to be the linear affine function on any interval $[k, k+1] \subset \mathbb{R}, k \in \mathbb{Z}$, which satisfies $\mathrm{m}_0(k) = M_k$ for $|k| \leq 2$ and $\mathrm{m}_0(k) = 0$ for $|k| \geq 4$. Taking into consideration that $\mathrm{supp}\, \phi_j \subset [2^{j-1} \leq |x| \leq 2^{j+1}]$ for all $j \geq 1$ and that m_j are continuous functions with compact support, we obtain from Lemma 2.2.2 that

$$\left\| \sum_{k \in \mathbb{Z}} \phi_j(k) \left(2^{(r-s)j} M_k\right) \widehat{f}(k) e^{ikx} \right\|_{C(\mathbb{S},\mathbb{C})} = \left\| \sum_{k \in \mathbb{Z}} \phi_j(k) \mathrm{m}_j(k) \widehat{f}(k) e^{ikx} \right\|_{C(\mathbb{S},\mathbb{C})}$$

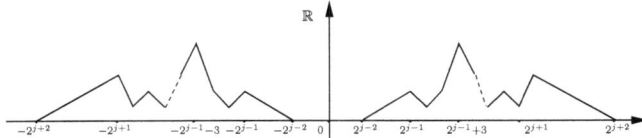

Figure 2.4: The function $\mathfrak{m}_j$, $j \geq 1$.

$$\leq (2\pi)^{-1/2}\|\mathcal{F}^{-1}\mathfrak{m}_j\|_{L^1} \left\|\sum_{k\in\mathbb{Z}}\phi_j(k)\widehat{f}(k)e^{ikx}\right\|_{C(\mathbb{S},\mathbb{C})}.$$

The estimate is valid also when $j = 0$. Hence, it suffices to prove that the inverse Fourier transforms $\mathcal{F}^{-1}\mathfrak{m}_j$ are uniformly bounded in $L^1(\mathbb{R},\mathbb{C})$, i.e.

$$\sup_{j\in\mathbb{N}}\|\mathcal{F}^{-1}\mathfrak{m}_j\|_{L^1} < \infty. \tag{2.37}$$

Since $s_1 < \infty$, we obtain for $j \in \mathbb{N}$ with $j \geq 1$ that

$$\|\mathfrak{m}_j\|_{C(\mathbb{R},\mathbb{C})} = \max_{2^{j-1}\leq k\leq 2^{j+1}} 2^{(r-s)j}|M_k| \leq s_1 \max_{2^{j-1}\leq k\leq 2^{j+1}} \frac{2^{(r-s)j}}{k^{r-s}}$$

$$\leq s_1 \left\{\begin{array}{ll} \frac{2^{(r-s)j}}{2^{(j-1)(r-s)}} & \text{if } r-s \geq 0; \\ \frac{2^{(r-s)j}}{2^{(j+1)(r-s)}} & \text{if } r-s < 0; \end{array}\right\} = s_1 2^{|r-s|},$$

so that $(\mathfrak{m}_j)_{j\geq 1}$ is uniformly bounded in $C(\mathbb{R},\mathbb{C})$, i.e.

$$\sup_{j\geq 1}\|\mathfrak{m}_j\|_{C(\mathbb{R},\mathbb{C})} \leq s_1 2^{|r-s|}. \tag{2.38}$$

Given $j \in \mathbb{N}, j \geq 1$, we define $\mathfrak{n}_j : \mathbb{R} \to \mathbb{C}$ by the relation $\mathfrak{n}_j(x) = \mathfrak{m}_j(2^j x)$ for all $x \in \mathbb{R}$. We observe that $\operatorname{supp}\mathfrak{n}_j \subset [4^{-1} \leq |x| \leq 4]$ and that relation (2.38) is valid for the sequence $(\mathfrak{n}_j)_{j\geq 1}$ too. Moreover,

$$(2\pi)^{1/2}\|\mathcal{F}^{-1}\mathfrak{n}_j\|_{C(\mathbb{R},\mathbb{C})} \leq \int_{|x|\leq 4}|\mathfrak{n}_j(x)|\,dx \leq 8\cdot 2^{|r-s|}s_1 \tag{2.39}$$

for all $j \geq 1$. Using the properties of the Fourier transform we obtain that $\mathcal{F}^{-1}\mathsf{n}_j(x) = 2^{-j}(\mathcal{F}^{-1}\mathsf{m}_j)(2^{-j}x)$ for all $j \geq 1$ and $x \in \mathbb{R}$, so that the variable substitution $x = 2^j y$ leads to

$$\|\mathcal{F}^{-1}\mathsf{n}_j\|_{L^1} = \int_{\mathbb{R}} |\mathcal{F}^{-1}\mathsf{n}_j|(x)\, dx = \int_{\mathbb{R}} 2^j |\mathcal{F}^{-1}\mathsf{n}_j|(2^j y)\, dy = \int_{\mathbb{R}} |\mathcal{F}^{-1}\mathsf{m}_j|(y)\, dy$$
$$= \|\mathcal{F}^{-1}\mathsf{m}_j\|_{L^1}.$$

Thus, instead of showing (2.37) we prove that the sequence $(\mathsf{n}_j)_{j \geq 1}$ is uniformly bounded in $L^1(\mathbb{R}, \mathbb{C})$ and that $\mathsf{m}_0 \in L^1(\mathbb{R}, \mathbb{C})$.

Integrating by parts twice, we obtain for $x \in \mathbb{R} \setminus \{0\}$ that

$$(2\pi)^{1/2}(\mathcal{F}^{-1}\mathsf{n}_j)(x) = 2^{-j} \int_{2^{j-2} \leq |x| \leq 2^{j+2}} \mathsf{m}_j(y) e^{i 2^{-j} y x}\, dy$$

$$= 4 \cdot 2^{(r-s)j} M_{2^{j-1}} \frac{e^{ix/2} - e^{ix/4}}{x^2} - \frac{1}{2} \cdot 2^{(r-s)j} M_{2^{j+1}} \frac{e^{i4x} - e^{i2x}}{x^2}$$

$$+ \frac{1}{2} \cdot 2^{(r-s)j} M_{-2^{j+1}} \frac{e^{-i2x} - e^{-i4x}}{x^2} - 4 \cdot 2^{(r-s)j} M_{-2^{j-1}} \frac{e^{-ix/4} - e^{ix/2}}{x^2}$$

$$+ \sum_{k=2^{j-1}}^{2^{j+1}-1} 2^j (2^{(r-s)j} M_{k+1} - 2^{(r-s)j} M_k) \frac{e^{i 2^{-j}(k+1)x} - e^{i 2^{-j} k x}}{x^2}$$

$$+ \sum_{k=-2^{j+1}}^{-2^{j-1}-1} 2^j (2^{(r-s)j} M_{k+1} - 2^{(r-s)j} M_k) \frac{e^{i 2^{-j}(k+1)x} - e^{i 2^{-j} k x}}{x^2}.$$

For $x \neq 0$, the sum of the first four terms from the right hand side of equality above can be bounded by $18 \cdot 2^{|r-s|} s_1 |x|^{-2}$ uniformly in $j \geq 1$, since

$$2^{(r-s)j} |M_{\pm 2^{j\pm 1}}| = \frac{2^{(r-s)j}}{2^{(r-s)(j\pm 1)}} | \pm 2^{(j\pm 1)} |^{r-s} |M_{\pm 2^{j\pm 1}}| \leq 2^{|r-s|} s_1$$

holds for all $j \in \mathbb{N}$ with $j \geq 1$. Rearranging, we have

$$\sum_{k=2^{j-1}}^{2^{j+1}-1} 2^j (2^{(r-s)j} M_{k+1} - 2^{(r-s)j} M_k) \left(e^{i 2^{-j}(k+1)x} - e^{i 2^{-j} k x} \right)$$

$$= -\sum_{p=2^{j-1}}^{2^{j+2}-2} 2^j 2^{(r-s)j} \left(M_{p+2} - 2M_{p+1} + M_p\right) e^{i2^{-j}(p+1)x}$$
$$+ 2^j 2^{(r-s)j}(M_{2^{j+1}} - M_{2^{j+1}-1})e^{i2x} - 2^j 2^{(r-s)j}(M_{2^{j-1}+1} - M_{2^{j-1}})e^{ix/2},$$

and, using estimates similar to those above, we find that

$$2^j 2^{(r-s)j}|M_{2^{j-1}+1} - M_{2^{j-1}}| \leq \frac{2^{(r-s+1)j}}{(2^{j-1})^{r-s+1}}|2^{j-1}|^{r-s+1}|M_{2^{j-1}+1} - M_{2^{j-1}}|$$
$$\leq 2^{|r-s+1|}s_2,$$

and analogously

$$2^j 2^{(r-s)j}|M_{2^{j+1}} - M_{2^{j+1}-1}| \leq \frac{2^{(r-s+1)j}}{(2^{j+1}-1)^{r-s+1}}s_2 \leq 2^{|r-s+1|}s_2.$$

Moreover,

$$\sum_{p=2^{j-1}}^{2^{j+2}-2} 2^j 2^{(r-s)j}|M_{p+2} - 2M_{p+1} + M_p| = 2^{(r-s+1)j}\sum_{p=2^{j-1}}^{2^{j+2}-2}|M_{p+2} - 2M_{p+1} + M_p|$$
$$= 2^{(r-s+1)j}\sum_{p=2^{j-1}}^{2^{j+2}-2}\frac{1}{p^{r-s+2}}p^{r-s+2}|M_{p+2} - 2M_{p+1} + M_p|$$
$$\leq 2^{(r-s+1)j}s_3\sum_{p=2^{j-1}}^{2^{j+2}-2}\frac{1}{p^{r-s+2}} \leq 2^{|r-s+2|}s_3.$$

The second sum of the expression for $(2\pi)^{-1/2}(\mathcal{F}^{-1}\mathfrak{n}_j)(x)$, when $x \neq 0$, can be estimated similarly. Summarising, we found for $x \neq 0$ that

$$|(2\pi)^{1/2}\mathcal{F}^{-1}(\mathfrak{n}_j)(x)| \leq \left(18 \cdot 2^{|r-s|}s_1 + 4 \cdot 2^{|r-s+1|}s_2 + 2 \cdot 2^{|r-s+2|}s_3\right)|x|^{-2},$$

and in view of (2.39) we found a uniform bound of the $L^1(\mathbb{R},\mathbb{C})$–norm of $\mathcal{F}^{-1}\mathfrak{n}_j$ when $j \geq 1$. It remains to show that $\mathcal{F}^{-1}\mathfrak{m}_0$ belongs to $L^1(\mathbb{R},\mathbb{C})$. The mapping $\mathfrak{m}_0$ is defined by

$$\mathfrak{m}_0(x) = \begin{cases} M_{-2}(4+x)/2 & , \quad -4 \leq x \leq -2, \\ M_{k+1}(x-k) + M_k(k+1-x) & , \quad x \in [k, k+1], -2 \leq k \leq 1, \\ M_2(4-x)/2 & , \quad 2 \leq t \leq 4, \end{cases}$$

so that for $x \neq 0$ we compute

$$(2\pi)^{-1/2}\mathcal{F}^{-1}\mathrm{m}_0(x) = \frac{M_{-2}}{2}\frac{e^{-i2x} - e^{-i4x}}{x^2} + \sum_{k=-2}^{1}(M_{k+1} - M_k)\frac{e^{i(k+1)x} - e^{ikx}}{x^2}$$

$$-\frac{M_2}{2}\frac{e^{i4x} - e^{i2x}}{x^2}.$$

Taking into consideration the following estimates

$$(2\pi)^{1/2}|\mathcal{F}^{-1}\mathrm{m}_0(x)| \leq (18\max_{-2\leq k\leq 2}|M_k|)\,x^{-2} \quad \text{for} \quad x \neq 0,$$
$$(2\pi)^{1/2}|\mathcal{F}^{-1}\mathrm{m}_0(x)| \leq 8\max_{-2\leq k\leq 2}|M_k| \quad \text{for all} \quad x \in \mathbb{R},$$

we obtain the desired result.

□

The Hölder spaces are compactly embedded, $C^s(\mathbb{S}) \hookrightarrow C^r(\mathbb{S})$ for $r < s$. However, the embedding is not dense if $r \notin \mathbb{N}$. In the situation when we deal with semigroups in the Hölder spaces context, it is natural to consider the closure of smooth functions in the classical Hölder spaces. More precisely, given $m \in \mathbb{N}$ and $\beta \in (0,1)$, we introduce the so-called small Hölder spaces $h^{m+\beta}(\mathbb{S})$ which are defined as the closure of the smooth functions $C^\infty(\mathbb{S})$ in the $C^{m+\beta}(\mathbb{S})$–norm. Since $C^\infty(\mathbb{S}) \subset h^{m+\beta}(\mathbb{S})$ for all $m \in \mathbb{N}$ and $\beta \in (0,1)$, the embedding $h^{m_1+\beta_1}(\mathbb{S}) \hookrightarrow h^{m_2+\beta_2}(\mathbb{S})$ is dense and compact whenever $m_2 + \beta_2 < m_1 + \beta_1$. In fact, the small Hölder spaces can be described intrinsically in the following way:

Lemma 2.2.3 *Let $f \in C^{m+\beta}(\mathbb{S})$ be given. Then*

$$f \in h^{m+\beta}(\mathbb{S}) \quad \Leftrightarrow \quad \lim_{\tau \searrow 0}\sup_{0<|x-x'|<\tau}\frac{|f^{(m)}(x) - f^{(m)}(x')|}{|x-x'|^\beta} = 0.$$

Proof We begin by showing the first implication. Let $\varepsilon > 0$ and $f \in h^{m+\beta}(\mathbb{S})$ be given. By the definition of the small Hölder spaces there exists a sequence $(f_n)_n \subset C^\infty(\mathbb{S})$ converging to f in $C^{m+\beta}(\mathbb{S})$. Choose $N \in \mathbb{N}$ such that $\|f_N - f\|_{C^{m+\beta}(\mathbb{S})} \leq \varepsilon/2$.

For $\tau < \bigl(\varepsilon/2\|f_N\|_{C^{m+1}(\mathbb{S})}\bigr)^{1/(1-\beta)}$ and $0 < |x - x'| < \tau$, we have that

$$\frac{|f^{(m)}(x) - f^{(m)}(x')|}{|x-x'|^\beta} \leq \|f - h_N\|_{C^{m+\beta}(\mathbb{S})} + \frac{|f_N^{(m)}(x) - f_N^{(m)}(x')|}{|x-x'|^\beta}$$

$$\leq \frac{\varepsilon}{2} + \|f_N^{(m+1)}\|_{C(\mathbb{S})} |x-x'|^{1-\beta} \leq \varepsilon.$$

Assume now that $f \in C^{m+\beta}(\mathbb{S})$, and additionally

$$\lim_{\tau \searrow 0} \sup_{0 < |x-x'| < \tau} \frac{|f^{(m)}(x) - f^{(m)}(x')|}{|x-x'|^\beta} = 0. \qquad (2.40)$$

Consider a smooth function $\varphi \in C^\infty(\mathbb{R})$ such that

$$\varphi \geq 0, \qquad \operatorname{supp} \varphi \subset (-1,1), \qquad \int_\mathbb{R} \varphi\, dx = 1,$$

and let $\varphi_\varepsilon(\cdot) := \varepsilon^{-1} \varphi(\cdot/\varepsilon)$, $\varepsilon \in (0,\infty)$, the mollifier corresponding to φ.

By considering a partition of unity of $\mathbb{S}$, we find cf. [5, Proposition 1.20], functions $\pi_j \in C^\infty(\mathbb{S})$ and compact subsets $K_j \subset (j\pi, (2+j)\pi)$ such that $\operatorname{supp} \pi_j = \cup_{k\in\mathbb{Z}}(2k\pi + K_j)$ and $\pi_1 + \pi_2 = 1$ on $\mathbb{S}$. Let f_j, $1 \leq j \leq 2$, be the function defined by $f_j := \chi_{(j\pi,(2+j)\pi)} \pi_j f$, where $\chi_{(j\pi,(2+j)\pi)}$ is the characteristic function of the interval $(j\pi, (2+j)\pi)$. In this way we have that

$$(f\pi_j)(x) = \sum_{k \in \mathbb{Z}} f_j(x + 2k\pi)$$

for all $x \in \mathbb{R}$ and $1 \leq j \leq 2$.

Given $\varepsilon \geq 0$ and $1 \leq j \leq 2$, the convolution $\varphi_\varepsilon * f_j$ is, cf. [5, Theorem 7.8], smooth and we infer from [5, Theorem 7.10] that $\operatorname{supp} \varphi_\varepsilon * f_j \subset (-\varepsilon, \varepsilon) + K_j \subset (j\pi, (2+j)\pi)$ for ε small enough. Moreover, $\varphi_\varepsilon * f_j \to_{\varepsilon \to 0} f_j$ in $BUC^m(\mathbb{R})$, the space consisting of functions with bounded and uniformly continuous derivatives up to order m. Hence, if ε is small enough, the extension

$$g_{j,\varepsilon}(\cdot) := \sum_{k\in\mathbb{Z}} \varphi_\varepsilon * f_j(\cdot + 2k\pi), \quad 1 \leq j \leq 2,$$

belongs to $C^\infty(\mathbb{S})$. Even more, given $j \in \{1,2\}$, we have that

$$g_{j,\varepsilon} - f\pi_j = \varphi_\varepsilon * f_j - f_j \quad \text{on} \quad (j\pi, (2+j)\pi). \qquad (2.41)$$

Thus, $g_{j,\varepsilon} \to f\pi_j$ in $C^m(\mathbb{S})$. We are left to prove that $g_{j,\varepsilon}^{(m)} \to_{\varepsilon \to 0} (f\pi_j)^{(m)}$ in $C^\beta(\mathbb{S})$. By (2.41), it suffices to show that $(\varphi_\varepsilon * f_j)^{(m)} \to_{\varepsilon \to 0} f_j^{(m)}$ in $\mathit{BUC}^\beta(\mathbb{R})$.

Let $\delta > 0$ be given. Our assumption (2.40) yields the existence of a constant $\vartheta > 0$ such that

$$\sup_{0 < |x-x'| < \vartheta} \frac{|f_j^{(m)}(x) - f_j^{(m)}(x')|}{|x - x'|^\beta} \leq \delta/2.$$

We may chose $\varepsilon_0 > 0$ such that

$$\|(\varphi_\varepsilon * f_j)^{(m)} - f_j^{(m)}\|_{BUC^m(\mathbb{R})} \leq \vartheta^\beta \delta/2$$

for all $\varepsilon \leq \varepsilon_0$. Let now $x, x' \in \mathbb{R}$ with $x \neq x'$. Using the relation $(\varphi_\varepsilon * f_j)^{(m)} = \varphi_\varepsilon * f_j^{(m)}$, we find for $|x - x'| \geq \vartheta$ that

$$\frac{|(\varphi_\varepsilon * f_j)^{(m)}(x) - f_j^{(m)}(x) - (\varphi_\varepsilon * f_j)^{(m)}(x') - f_j^{(m)}(x')|}{|x - x'|^\beta}$$

$$\leq \frac{|\varphi_\varepsilon * f_j^{(m)}(x) - f_j^{(m)}(x)|}{\vartheta^\beta} + \frac{|\varphi_\varepsilon * f_j^{(m)}(x') - f_j^{(m)}(x')|}{\vartheta^\beta} \leq \delta.$$

If $|x - x'| < \vartheta$ we have that

$$\frac{|(\varphi_\varepsilon * f_j)^{(m)}(x) - f_j^{(m)}(x) - (\varphi_\varepsilon * f_j)^{(m)}(x') - f_j^{(m)}(x')|}{|x - x'|^\beta}$$

$$\leq \frac{|f_j^{(m)}(x) - f_j^{(m)}(x')|}{|x - x'|^\beta} + \frac{|\varphi_\varepsilon * f_j^{(m)}(x) - \varphi_\varepsilon * f_j^{(m)}(x')|}{|x - x'|^\beta}$$

$$\leq \frac{\delta}{2} + \int_\mathbb{R} \varepsilon^{-1} \varphi_\varepsilon(y/\varepsilon) \frac{|f_j^{(m)}(x - y) - f_j^{(m)}(x' - y)|}{|x - x'|^\beta} \, dy$$

$$\leq \frac{\delta}{2} + \frac{\delta}{2} \int_\mathbb{R} \varepsilon^{-1} \varphi_\varepsilon(y/\varepsilon) \, dy = \delta.$$

Hence $f_\varepsilon := g_{1,\varepsilon} + g_{2,\varepsilon} \in C^\infty(\mathbb{S})$, and $f_\varepsilon \to_{\varepsilon \to 0} f\pi_1 + f\pi_2 = f$ in $C^{m+\beta}(\mathbb{S})$. This completes the proof.

□

The small Hölder space $h^{m+\beta}(\mathbb{S})$ is a proper subspace of $C^{m+\beta}(\mathbb{S})$. A function f which in a neighbourhood of 0 is defined by

$$f(x) = \begin{cases} 0, & x \leq 0 \\ x^\beta, & x > 0, \end{cases}$$

can be extended to a function belonging to $C^\beta(\mathbb{S})$. However, in virtue of Lemma 2.2.3, $f \notin h^\beta(\mathbb{S})$.

We denote in the following by $(\cdot|\cdot)_\theta = (\cdot|\cdot)_{\theta,\infty}^0$ the interpolation functor introduce by Da Prato and Grisvard [19]. It is well-known that the small Hölder spaces have the following interpolation property

$$(h^{\sigma_0}(\mathbb{S}), h^{\sigma_1}(\mathbb{S}))_\theta = h^{(1-\theta)\sigma_0 + \theta\sigma_1}(\mathbb{S}), \tag{2.42}$$

if $\theta \in (0,1)$ and $(1-\theta)\sigma_0 + \theta\sigma_1 \notin \mathbb{N}$. This property is a very useful tool and will help us establish the local well-posedness result stated in Theorem 3.2.3. Also, it will help us to construct an initial approximation for the solution of the Muskat problem, cf. Lemma 7.1.5.

Part I

Well-posedness, stability, and bifurcation results for the flow of a ferrofluid in a rotating Hele–Shaw cell

Chapter 3

The mathematical model

3.1 The physical setting

We describe first the mathematical model we deal with. Consider two parallel same-sized plates with a small distance $b << 1$ to each other. In the infinitesimally thin space between the plates is a viscous incompressible ferrofluid, which at time $t \geq 0$, occupies the bounded region $\Omega(t)$ in the Euclidean space $\mathbb{R}^3$. Its boundary $\Gamma(t)$ is subjected to surface tension effects and describes the propagation of the fluid inside the horizontal Hele-Shaw cell. This fluid system rotates, with constant angular velocity ω, around a rotation axis (set to be the Oz-axis) and the centre of the cell is aligned with the axis of rotation (see Figure 3.1). The point $0 \in \mathbb{R}^3$ is chosen to be the centre of the cell. The fluid is surrounded by air at uniform pressure $p_A = 0$ and is treated as a generalised Newtonian fluid with viscosity $\mu \in C^\omega([0,\infty),(0,\infty))$ which depends only of the shear rate and which satisfies the general conditions (2.25).

In general, the viscosity of a ferrofluid depends not only on shear rate but also on temperature and on the magnetic field applied. Though, if the temperature and the magnetisation of the ferrofluid are constant, then the fluid shows a shear thinning characteristic, the viscosity decreases with increasing rate of shear (see [40, 44]). Also, if the ferrofluid is at saturation, meaning that a state is reached when an increase in applied external magnetising field $\vec{H}$ cannot increase the magnetisation of the material further, then the viscosity does not depend on the magnetisation. As we have seen in the second chapter of this work, the shear thinning fluids fit very

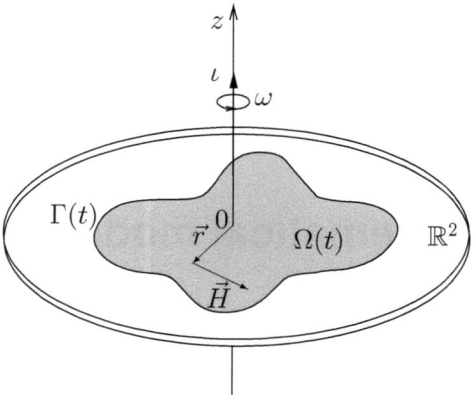

Figure 3.1: The rotating Hele-Shaw cell.

well in our context since the majority of the examples considered there presented this property.

To include magnetic forces, we consider the action of an external radial magnetic field $\vec{H}$, produced by a long, straight wire, which is also the rotation axis, carrying a current ι. By Amperes law, it can be shown that the steady current with intensity ι produces an azimuthal magnetic field external to the wire $\vec{H} = \iota/(2\pi r)\vec{\theta}$, where r is the distance from the wire, $\vec{\theta}$ is the unit vector $\vec{\theta} = i\vec{r}/r$ and $\vec{r}$ denotes the projection of the position vector $\vec{x}$ onto $\mathbb{R}^2$, the plane parallel to the plates which contains the point $0 \in \mathbb{R}^3$. We denoted by i the root of -1. Following the standard approximations used by Rosensweig [57] we assume that the ferrofluid's magnetisation $\vec{M}$ is collinear with the external field $\vec{H}$ and that the influence of the demagnetising field is neglected. The external forces acting on the fluid body are the magnetic force $\mathfrak{f}_m$, which cf. (2.16) is equal to $\mathfrak{f}_m = \mu_0 (\vec{M} \cdot \nabla)\vec{H}$ and the centrifugal force

$$\mathfrak{f}_c = \rho \omega^2 \vec{r}.$$

Due to the small distance between the plates, the inertial forces (including the Coriolis force) on the fluid are small compared to the viscosity. In our

situation, the magnetic force is normal to the Oz−axis and is given by
$$\mathfrak{f}_m = \nabla\left(\frac{\mu_0\chi H^2}{2}\right) = -\frac{\mu_0\chi\iota^2}{(2\pi)^2}\cdot\frac{\vec{r}}{r^4},$$
since $M = \chi H$, with χ a constant magnetic susceptibility. To simplify our notation we set $K := \mu_0\chi/(8\pi^2)$. Defining the potential u by the relation
$$u = p - K\iota^2\frac{1}{r^2} - \frac{\rho\omega^2}{2}r^2, \tag{3.1}$$
we find out that relation (2.17) is satisfied and we are in the situation considered in the first section of Chapter 2. Hence, by integrating over the small gap between the plates, we come to a two-dimensional setting. By (2.29), the gap averaged velocity $\vec{v}$ satisfies then
$$\vec{v} = -\frac{\nabla u}{\overline{\mu}(|\nabla u|^2)} \quad \text{in} \quad \Omega(t), \tag{3.2a}$$
where $\overline{\mu}$ is defined by (2.30), whereas, from the conservation of mass (2.32), we get
$$\nabla\vec{v} = 0 \quad \text{in} \quad \Omega(t). \tag{3.2b}$$
We still write $\Omega(t)$ and $\Gamma(t)$ for the projection of $\Omega(t)$ and $\Gamma(t)$ onto $\mathbb{R}^2$, respectively. For unit density $\rho = 1$, the Laplace-Young condition (2.33) leads to
$$u = \gamma\kappa_{\Gamma(t)} - K\iota^2\frac{1}{r^2} - \frac{\omega^2}{2}r^2 \quad \text{on} \quad \Gamma(t), \tag{3.2c}$$
where $\kappa_{\Gamma(t)}$ stands for the curvature of $\Gamma(t)$. A second boundary condition is then given by
$$V(t,\cdot) = \langle\vec{v}(t,\cdot),\nu(t)\rangle \quad \text{on} \quad \Gamma(t), \tag{3.2d}$$
meaning that the interface moves along with the fluid (a fluid particle on the boundary remains on the boundary as time elapses). We let $\nu(t)$ denote the outward unit normal at $\Omega(t)$. The fluid domain at time $t = 0$ is considered to be known, so that
$$\Omega(0) = \Omega_0. \tag{3.2e}$$
The equations (3.2) constitute a single-phase moving boundary problem. The main goal is to determine the motion of a interface separating the fluid from air which at time $t = 0$ is known and also the potential u inside the fluid blob. Particularly, we aim to find the steady-state solutions of the problem and to study their stability properties.

3.2 The mathematical model

We are interested first in proving that the problem (3.2) is locally well-posed in time. To this scope we rewrite (3.2) in a more accessible way. By identifying functions over the unit circle $\mathbb{S}$ with 2π−periodic functions on the real line we consider the Hölder spaces and the small Hölder spaces $C^{k+\beta}(\mathbb{S})$ and $h^{k+\beta}(\mathbb{S})$, $k \in \mathbb{N}, \beta \in (0,1)$, respectively, defined in the second section of the Chapter 2.

For our further analysis we fix $\alpha \in (0,1)$, and set

$$\mathcal{V} := \begin{cases} \{\varrho \in h^{4+\alpha}(\mathbb{S}) : \|\varrho\|_{C(\mathbb{S})} < 1/4\}, & \text{if (2.26) hold;} \\ \{\varrho \in h^{4+\alpha}(\mathbb{S}) : \|\varrho\|_{C^2(\mathbb{S})} < 1/8\}, & \text{else.} \end{cases}$$

We assumed from the beginning that the viscosity function μ satisfies the relations (2.25). We stress that $\mathcal{V}$ is an open neighbourhood of the origin in $h^{4+\alpha}(\mathbb{S})$. Given $\varrho \in \mathcal{V}$, we define the simple connected domain

$$\Omega_\varrho := \left\{ y \in \mathbb{R}^2 : |y| < 1 + \varrho\left(\frac{y}{|y|}\right) \right\} \cup \{0\},$$

with boundary $\partial \Omega_\varrho =: \Gamma_\varrho$, expressed by

$$\Gamma_\varrho = \left\{ y \in \mathbb{R}^2 : |y| = 1 + \varrho\left(\frac{y}{|y|}\right) \right\} = \{(1 + \varrho(x))x : x \in \mathbb{S}\}.$$

In the remainder of this part we steadily use y to denote the position vector $y \in \mathbb{R}^2$, while x stands for the variable on $\mathbb{S}$. It is suitable to represent Γ_ϱ as the 0−level set of an appropriate function. For this, let $N_\varrho : R(3/4, 5/4) \to \mathbb{R}$ be the function defined by

$$N_\varrho(y) = |y| - 1 - \varrho(y/|y|),$$

where $R(3/4, 5/4)$ is the circular ring centred in 0 with radii $3/4$ and $5/4$, i.e.

$$R(3/4, 5/4) := \{y \in \mathbb{R}^2 : 3/4 < |y| < 5/4\}.$$

Since $\Gamma_\varrho = N_\varrho^{-1}(0)$ it follows that the outward normal ν_ϱ at Γ_ϱ is

$$\nu_\varrho = \nabla N_\varrho / |\nabla N_\varrho|,$$

with ∇N_ϱ denoting the gradient of N_ϱ. A simple computation shows, cf. [23], that for $y \neq 0$,

$$\nabla\left(\varrho\left(\frac{y}{|y|}\right)\right) = \varrho'\left(\frac{y}{|y|}\right)\left(-\frac{y_2}{|y|^2}, \frac{y_1}{|y|^2}\right), \tag{3.3}$$

hence

$$\nabla N_\varrho(\Theta_\varrho(x)) = x - \frac{\varrho'(x)}{1+\varrho(x)}(-x_2, x_1) \tag{3.4}$$

for all $x \in \mathbb{S}$. If one of the conditions (2.26) is not satisfied, then Ω_ϱ is convex for all $\varrho \in \mathcal{V}$ and Γ_ϱ has strictly positive curvature. Indeed, for $\varrho \in \mathcal{V}$, a straightforward calculation gives

$$\kappa_{\Gamma_\varrho}(y) = \frac{(1+\varrho)^2 + 2\varrho'^2 - (1+\varrho)\varrho''}{((1+\varrho)^2 + \varrho'^2)^{3/2}}\left(\frac{y}{|y|}\right), \quad y \in \Gamma_\varrho, \tag{3.5}$$

hence

$$\begin{aligned}
|\kappa_{\Gamma_\varrho} - 1| &= \left|\frac{(1+\varrho)^2 + 2\varrho'^2 - (1+\varrho)\varrho''}{((1+\varrho)^2 + \varrho'^2)^{3/2}} - 1\right| \leq \left|\frac{(1+\varrho)^2}{((1+\varrho)^2 + \varrho'^2)^{3/2}} - 1\right| \\
&\quad + \frac{2\varrho'^2}{((1+\varrho)^2 + \varrho'^2)^{3/2}} + \frac{(1+\varrho)|\varrho''|}{((1+\varrho)^2 + \varrho'^2)^{3/2}} \leq \frac{2/8^2}{(1-1/8)^3} \\
&\quad + \frac{(1+1/8)1/8}{(1-1/8)^3} + \frac{|(1+\varrho)^2 - ((1+\varrho)^2 + \varrho'^2)^{3/2}|}{((1+\varrho)^2 + \varrho'^2)^{3/2}} \\
&\leq \frac{|(1+\varrho)^4 - ((1+\varrho)^2 + \varrho'^2)^3|}{(1-1/8)^3((1-1/8)^2 + (1-1/8)^3)} + \frac{3/8^2 + 1/8}{(1-1/8)^3} \\
&\leq 0.8.
\end{aligned}$$

Consequently $\kappa_{\Gamma_\varrho} \geq 0.2$ and the domain Ω_ϱ is strictly convex. The strict convexity of Ω_ϱ is needed to guarantee the solvability of some quasilinear elliptic Dirichlet problem associated to our model (3.2).

To incorporate time let $T > 0$. Assuming the mapping $\varrho \in C([0,T], \mathcal{V}) \cap C^1([0,T], h^{1+\alpha}(\mathbb{S}))$ describes the evolution of the bulk fluid which occupies initially the domain Ω_0, that is $\Omega(t) = \Omega_{\varrho(t)}$, $0 \leq t \leq T$, then (3.2d) is equivalent to

$$V(t, y) = -\frac{\partial_t N_{\varrho(t)}(y)}{|\nabla N_{\varrho(t)}(y)|} = \frac{\partial_t \varrho(t)(y/|y|)}{|\nabla N_{\varrho(t)}(y)|}, \quad y \in \Gamma_{\varrho(t)},$$

which together with (3.2) leads to the following system

$$\begin{cases} \mathcal{Q}u = 0 & \text{in } \Omega_{\varrho(t)}, \ 0 \leq t \leq T, \\ u = \gamma \kappa_{\Gamma_{\varrho(t)}} - K\iota^2 \dfrac{1}{|y|^2} - \dfrac{\omega^2}{2}|y|^2 & \text{on } \Gamma_{\varrho(t)}, \ 0 \leq t \leq T, \\ \partial_t N_\varrho = \dfrac{1}{\overline{\mu}(|\nabla u|^2)}\langle \nabla u | \nabla N_\varrho \rangle & \text{on } \Gamma_{\varrho(t)}, \ 0 \leq t \leq T, \\ \varrho(0) = \varrho_0, \end{cases} \quad (3.6)$$

where $\mathcal{Q}u = \operatorname{div}(\nabla u / \overline{\mu}(|\nabla u|^2))$. Using the standard sum convention

$$a_{ij}(\nabla u) u_{ij} = \sum_{i,j=1}^{2} a_{ij}(\nabla u) u_{ij}$$

we compute that the quasilinear operator $\mathcal{Q}$ is given by $\mathcal{Q}u = a_{ij}(\nabla u)u_{ij}$, where

$$a_{ij}(p) = \frac{\delta_{ij}}{\overline{\mu}(|p|^2)} - \frac{2 p_i p_j \overline{\mu}'(|p|^2)}{\overline{\mu}^2(|p|^2)}, \quad 1 \leq i, j \leq 2 \quad (3.7)$$

for all $p = (p_1, p_2) \in \mathbb{R}^2$. Under the assumptions (2.25), $\mathcal{Q}$ is a quasilinear elliptic operator. More precisely, we have:

Lemma 3.2.1 *The operator $\mathcal{Q}$ is a quasilinear elliptic operator in $\mathbb{R}^2$. Given $p \in \mathbb{R}^2$, there exist positive constants $\lambda(p), \Lambda(p)$, such that*

$$\lambda(p)|\xi|^2 \leq a_{ij}(p)\xi_i \xi_j \leq \Lambda(p)|\xi|^2 \quad \text{for all} \quad \xi = (\xi_1, \xi_2) \in \mathbb{R}^2.$$

Moreover, if (2.26) are satisfied, then $\mathcal{Q}$ is uniformly elliptic, i.e. λ and Λ can be chosen independently of $p \in \mathbb{R}^2$.

Proof Let $\lambda(p) := \min\{\lambda_1(p), \lambda_2(p)\}$ and $\Lambda(p) := \max\{\lambda_1(p), \lambda_2(p)\}$ for $p \in \mathbb{R}^2$, where $\lambda_i(p), 1 \leq i \leq 2$, the eigenvalues of the matrix $[a_{ij}(p)]$, are

$$\lambda_1(p) = \frac{1}{\overline{\mu}(|p|^2)} \quad \text{and} \quad \lambda_2(p) = \frac{1}{\overline{\mu}(|p|^2)} - \frac{2|p|^2 \overline{\mu}'(|p|^2)}{\overline{\mu}^2(|p|^2)}.$$

We are left to prove that λ and Λ are positive, and uniformly bounded from above and below if (2.26) are satisfied. Recall that

$$\frac{1}{\overline{\mu}(r)} = \overline{c} \int_{-\overline{b}}^{\overline{b}} \frac{s^2}{\overline{\mu}(s^2 r)} \, ds \quad \text{for } r \geq 0,$$

where $\widetilde{\mu} := \mu \circ h^{-1}$ and $h(r) := r\mu^2(r)$. Since μ is positive, then so are $\widetilde{\mu}$ and $\overline{\mu}$, hence λ_1 is positive. Moreover, if (2.26) are satisfied, then λ_1 is bounded from above and below uniformly in $p \in \mathbb{R}^2$

$$\frac{2\overline{c}b^3}{3M_\mu} \leq \frac{1}{\overline{\mu}(r)} \leq \frac{2\overline{c}b^3}{3m_\mu} \quad \text{for } r \geq 0,$$

which implies the uniform boundedness of λ_1. We now study the second eigenvalue λ_2. Differentiation yields

$$\frac{1}{\overline{\mu}(r)} - \frac{2r\overline{\mu}'(r)}{\overline{\mu}^2(r)} = \frac{1}{\overline{\mu}(r)} + 2r\left(\frac{1}{\overline{\mu}}\right)'(r)$$

$$= \overline{c}\int_{-\overline{b}}^{\overline{b}} \left[\frac{s^2}{\widetilde{\mu}(s^2r)} - \frac{2rs^4\widetilde{\mu}(s^2r)}{\widetilde{\mu}_-^2(s^2r)}\right] ds$$

$$= \overline{c}\int_{-\overline{b}}^{\overline{b}} s^2 \left[\frac{1}{\widetilde{\mu}(s^2r)} - \frac{2(s^2r)\widetilde{\mu}(s^2r)}{\widetilde{\mu}_-^2(s^2r)}\right] ds.$$

Consequently, it suffices to prove that the mapping

$$\left[[0, \infty) \ni r \mapsto \frac{1}{\widetilde{\mu}(r)} - \frac{2r\widetilde{\mu}'(r)}{\widetilde{\mu}^2(r)}\right]$$

is positive and uniformly bonded from below and above if (2.26) hold ture. Given $r \geq 0$, we set $s := h^{-1}(r)$ and obtain that

$$\frac{1}{\widetilde{\mu}(r)} - \frac{2r\widetilde{\mu}'(r)}{\widetilde{\mu}^2(r)} = \frac{1}{\mu^2(h^{-1}(r))}\left(\mu(h^{-1}(r)) - 2r\mu'(h^{-1}(r))(h^{-1})'(r)\right)$$

$$= \frac{1}{\mu^2(s)}\left(\mu(s) - 2h(s)\mu'(s)\frac{1}{h'(s)}\right)$$

$$= \frac{1}{\mu^2(s)}\left(\mu(s) - \frac{2s\mu^2(s)\mu'(s)}{\mu^2(s) + 2s\mu(s)\mu'(s)}\right)$$

$$= \frac{1}{\mu(s) + 2s\mu'(s)}.$$

The conclusion is now a simple consequence of relations (2.25) and (2.26).
□

It is of interest to determine the regularity properties of the coefficients a_{ij}, $1 \leq i, j \leq 2$, of the quasilinear operator $\mathcal{Q}$. Since $\bar{\mu} > 0$, this reduces to the study of the regularity of the effective viscosity function $\bar{\mu}$.

Lemma 3.2.2 *The effective viscosity $\bar{\mu}$ defined by (2.30) is a real analytic function on $[0, \infty)$, i.e. $\bar{\mu} \in C^\omega([0, \infty), (0, \infty))$.*

Proof Let us first observe that $h \in C^\omega([0, \infty), [0, \infty))$. According to (2.25) $h'(r) = \mu(r)(\mu(r) + r\mu'(r)) > 0$, so that $h^{-1} \in C^\omega([0, \infty), [0, \infty))$. Consequently, the composition $\tilde{\mu} = \mu \circ h^{-1}$ is also real analytic.

The variable substitution $t = s^2 r$ yields that

$$\bar{\mu}(r) = \frac{1}{\bar{c}} \frac{r^{3/2}}{\int_0^{\bar{b}^2 r} \sqrt{t} \tilde{\mu}(t)^{-1} \, dt},$$

and taking into consideration that $\tilde{\mu} > 0$, we get $\bar{\mu} \in C^\omega((0, \infty), (0, \infty))$.

We are thus left to prove that $\bar{\mu}$ is analytic in a neighbourhood of 0. Using the well-known analytic continuation of real analytic functions we can extend the restriction $\tilde{\mu}|[0, 1]$ to a holomorphic function f on some open neighbourhood $G \subset \mathbb{R}^2$ of $[0, 1]$. Even more, we can choose G to be convex and f bounded in G. Since $\tilde{\mu} > 0$ we can presuppose, in virtue of the identity theorem for holomorphic functions, that $\inf_G |f| > 0$.

Given $z \in G$, the path $[-\bar{b}, \bar{b}] \ni s \to s^2 z$ is contained in G, since $b < 1$ and G is convex. The function

$$[-\bar{b}, \bar{b}] \times G \ni (s, z) \mapsto \frac{s^2}{f(s^2 z)} \in \mathbb{C}$$

is therefore well-defined and satisfies the assumptions of the theorem on the differentiability of parameter integrals, cf. [5, Corollary 3.19]. Consequently, $\bar{\mu}|[0, 1]$ is, as restriction to $[0, 1]$ of the holomorphic function defined by integrating the function above on $[-\bar{b}, \bar{b}]$, real analytic. □

We introduce now the notion of solution of the moving boundary problem (3.6). We shall say that the pair (u, ϱ) is a classical Hölder solution of

(3.6) if
$$\varrho \in C([0,T], \mathcal{V}) \cap C^1([0,T], h^{1+\alpha}(\mathbb{S})),$$
$$u(\,\cdot\,,t) \in buc^{2+\alpha}(\Omega_{\varrho(t)}),\ t \in [0,T],$$

and if (u, ϱ) satisfies the equations in (3.6) pointwise.

Given $U \subset \mathbb{R}^2$ opened and $k \in \mathbb{N} \cup \{\infty\}$, the set $BUC^k(U)$ denotes the space of all maps form U to $\mathbb{R}$ which have bounded and uniformly continuous derivatives up to order k. The space $BUC^{k+\alpha}(U)$, $\beta \in (0,1)$, consists of all $u \in BUC^k(U)$ having uniformly α–Hölder continuous derivatives of order k. Finally, we set $buc^{k+\alpha}(U)$ to be the closure of $BUC^\infty(U)$ in $BUC^{k+\alpha}(U)$.

Let us notice that if (u, ϱ) is a solution of (3.6) and (2.26) are not satisfied, the fluid domain $\Omega_{\varrho(t)}$ must be convex for all $t \in [0,T]$. This is no longer the case when (2.26) is satisfied. The set $\mathcal{V}$ is large enough to contain functions that parametrise domains which are not convex. For example, the mapping ϱ given by $\varrho(x) = -(1/5)x_1^{10}$ for $x \in \mathbb{S}$ belongs to $\mathcal{V}$ and the curvature of Γ_ϱ in point $(1 - 1/5)$ is negative.

The first main result of this part is the following existence and uniqueness theorem:

Theorem 3.2.3 (Existence and uniqueness) *Let ω, γ and ι be two positive constants and $\alpha \in (0,1)$. There is an open neighbourhood $\mathcal{O} \subset h^{4+\alpha}(\mathbb{S})$ of the zero function such that for each initial value $\varrho_0 \in \mathcal{O}$, there exists $T > 0$ and a unique classical Hölder solution to problem (3.6) on $[0,T]$. Moreover, the mapping ϱ is analytical in the time variable on $(0,T)$.*

Having proved this theorem, we can immediately establish that the volume of ferrofluid is preserved by the flow (3.6). This property will play an important role when studying the stability properties of the circular equilibrium when the fluid blob occupies the unit disc (see the discussion preceding Theorem 5.1.2).

Observation 3.2.4 *The solutions of (3.6) preserve the volume of the fluid domain.*

Proof Indeed, taking into consideration that $\mathrm{vol}(\Omega_{\varrho(t)}) = \int_{\Omega_{\varrho(t)}} dy$ for all

$t \in [0, T]$, and using Stokes' theorem we have

$$\frac{d}{dt}\left(\int_{\Omega_{\varrho(t)}} dy\right) = \frac{d}{dt}\left(\frac{1}{2}\int_{\Gamma_{\varrho(t)}} \langle y, \nu_{\varrho(t)}\rangle\, ds\right) = \int_{\Gamma_{\varrho(t)}} V(t)\, ds$$

$$= -\int_{\Gamma_{\varrho(t)}} \left\langle \frac{\nabla u}{\bar{\mu}(|\nabla u|^2)} \bigg| \nu_{\varrho(t)} \right\rangle ds$$

$$= -\int_{\Omega_{\varrho(t)}} \nabla\left(\frac{\nabla u}{\bar{\mu}(|\nabla u|^2)}\right) dy = 0.$$

□

3.3 The transformed problem

If the evolution of the relative boundary ϱ is known, then we can solve the first two equations of system (3.6) as a boundary value problem for u, cf. Theorem 4.1.1. However, the third equation of (3.6) shows that we can determine ϱ only if we know u. We avoid this difficulty by transforming problem (3.6) onto the unitary disc $\mathbb{D} := D_{\mathbb{R}^2}(0, 1)$. To this scope, we define for each $\varrho \in \mathcal{V}$ the so-called Hanzawa diffeomorphism $\Theta_\varrho : \mathbb{R}^2 \to \mathbb{R}^2$ by the relation

$$\Theta_\varrho(y) = \begin{cases} y + \varphi(|y| - 1)\varrho\left(\frac{y}{|y|}\right)\frac{y}{|y|} & , \quad 0 < |y| < 2, \\ y & , \quad \text{else,} \end{cases}$$

where $\varphi \in C^\infty(\mathbb{R}, [0, 1])$ satisfies

$$\varphi(r) = \begin{cases} 1 & , \quad |r| \leq 1/4, \\ 0 & , \quad |r| \geq 3/4, \end{cases}$$

and additionally $\max_\mathbb{R} |\varphi'| < 1//4$. In fact for $|y| \leq 1/4$ or $|y| \geq 7/4$ we have that $||y| - 1| \geq 3/4$, so that $\Theta_\varrho(y) = y$. Given $y \in \mathbb{S}$, the mapping $[0, \infty) \ni r \mapsto r + \varphi(r-1)\varrho(y/|y|) \in [0, \infty)$ is strictly increasing and therefore bijective. Hence, being the perturbation of the identity $\text{id}_{\mathbb{R}^2}$ by a function

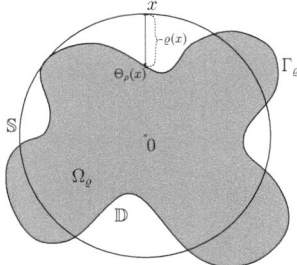

Figure 3.2: The Hanzawa diffeomorphism Θ_ϱ.

with compact support, we have that $\Theta_\varrho \in \textit{Diff}^{4+\alpha}(\mathbb{D}, \Omega_\varrho) \cap \textit{Diff}^{4+\alpha}(\mathbb{R}^2, \mathbb{R}^2)$. Additionally, Θ_ϱ maps the unit circle $\mathbb{S}$ onto Γ_ϱ, i.e. $\Theta_\varrho(\mathbb{S}) = \Gamma_\varrho$. The situation is illustrated in Figure 3.2.

For any $\varrho \in \mathcal{V}$, the Hanzawa diffeomorphism Θ_ϱ induces the pull-back and push-forward operators

$$\Theta_\varrho^* : \textit{BUC}(\Omega_\varrho) \to \textit{BUC}(\mathbb{D}), \quad u \mapsto u \circ \Theta_\varrho,$$

and

$$\Theta_*^\varrho : \textit{BUC}(\mathbb{D}) \to \textit{BUC}(\Omega_\varrho), \quad v \mapsto v \circ \Psi_\varrho,$$

respectively. We set here Ψ_ϱ to be the inverse of Θ_ϱ, i.e. $\Psi_\varrho := \Theta_\varrho^{-1}$.

From the definition of Θ_ϱ it is clear that, given $\varrho_0 \in \mathcal{V}$, there exists $\delta = \delta(\varrho_0) > 0$ and $C = C(\delta_0) > 0$ such that

$$\|\Theta_\varrho - \Theta_{\varrho_0}\|_{BUC^{4+\alpha}(\mathbb{R}^2)} \leq C \|\varrho - \varrho_0\|_{C^{4+\alpha}(\mathbb{S})} \tag{3.8}$$

for all $\|\varrho - \varrho_0\|_{C^{4+\alpha}(\mathbb{S})} \leq \delta$. The chain rule together with (3.3) yields for $0 < |y| < 2$ that

$$\psi_{\varrho,j}^i(\Theta_\varrho(y)) = \frac{\delta_{ij}}{1+\varphi'\varrho} + (-1)^{j'}\frac{(-1)^i y_{i'} y_{j'} \varphi\varrho + (-1)^{i'} y_{i'} y_{j'} |y|\varphi'\varrho + y_i y_{j'} \varphi\varrho'}{|y|^3 \left(1 + \frac{1}{|y|}\varphi\varrho\right)(1+\varphi'\varrho)}, \tag{3.9}$$

where $\varphi = \varphi(|y|-1)$, $\varrho = \varrho(y/|y|)$ and $\Psi_\varrho = (\psi_\varrho^1, \psi_\varrho^2)$. Given $i \in \{1,2\}$, we set

$$i' := \begin{cases} 2, & i=1, \\ 1, & i=2. \end{cases}$$

Consequently, we can choose $\delta = \delta(\varrho_0) > 0$ and $C = C(\delta_0) > 0$ in (3.8) such that

$$\|\Psi_\varrho - \Psi_{\varrho_0}\|_{BUC^{4+\alpha}(\mathbb{R}^2)} \leq C\|\varrho - \varrho_0\|_{C^{4+\alpha}(\mathbb{S})} \tag{3.10}$$

for all $\|\varrho - \varrho_0\|_{C^{4+\alpha}(\mathbb{S})} \leq \delta$.

To pull back the entire problem to $\mathbb{D}$, it is then natural to introduce the differential operator

$$\mathcal{A}(\varrho) := \Theta_\varrho^* \circ \mathcal{Q} \circ \Theta_*^\varrho, \qquad \varrho \in \mathcal{V}, \tag{3.11}$$

acting on $BUC^{2+\alpha}(\mathbb{D})$. Given $v \in BUC^{2+\alpha}(\mathbb{D})$, we have that

$$\mathcal{A}(\varrho)v = b_{ij}(y, \varrho, \nabla v)v_{ij} + b_i(y, \varrho, \nabla v)v_i,$$

where

$$b_{ij}(y, \varrho, \nabla v) = \psi_{\varrho,k}^i(\Theta_\varrho(y))\psi_{\varrho,l}^j(\Theta_\varrho(y))a_{kl}(\nabla(\Theta_*^\varrho v)(\Theta_\varrho(y))), \text{ for } 1 \leq i,j \leq 2, \tag{3.12}$$

$$b_i(y, \varrho, \nabla v) = \psi_{\varrho,kl}^i(\Theta_\varrho(y))a_{kl}(\nabla(\Theta_*^\varrho v)(\Theta_\varrho(y))), \text{ for } 1 \leq i \leq 2, \tag{3.13}$$

and

$$\nabla(\Theta_*^\varrho v)(\Theta_\varrho(y)) = (v_k(y)\psi_{\varrho,1}^k(\Theta_\varrho(y)), v_k(y)\psi_{\varrho,2}^k(\Theta_\varrho(y))) \tag{3.14}$$

for $v \in BUC^{2+\alpha}(\mathbb{D})$ and $y \in \mathbb{D}$. Moreover, given $\varrho \in \mathcal{V}$, $y \in \mathbb{D}$ and $v \in BUC^{2+\alpha}(\mathbb{D})$, we have that

$$b_{ij}(y, \varrho, v)\xi_i\xi_j = a_{ij}(\nabla(\Theta_*^\varrho v)(\Theta_\varrho(y)))(\xi_k\psi_{\varrho,i}^k(\Theta_\varrho(y)))(\xi_k\psi_{\varrho,j}^k(\Theta_\varrho(y)))$$

for all $\xi = (\xi_1, \xi_2) \in \mathbb{R}^2$. We infer from Lemma 3.2.1 that $\mathcal{A}(\varrho)$ is a quasi-linear elliptic operator. In particular, if the relations (2.26) are fulfilled, then $\mathcal{A}(\varrho)$ is uniformly elliptic.

To deal with the kinematic boundary condition of (3.6), we introduce for $(\varrho, v) \in \mathcal{V} \times BUC^{2+\alpha}(\mathbb{D})$ the boundary value $\mathcal{B}(\varrho, v) \in C^{1+\alpha}(\mathbb{S})$ by

$$\mathcal{B}(\varrho, v)(x) := -\frac{1}{\overline{\mu}(|\nabla(\Theta_*^\varrho v)|^2)} \langle \nabla(\Theta_*^\varrho v) | \nabla N_\varrho \rangle (\Theta_\varrho(x)), \quad x \in \mathbb{S}. \tag{3.15}$$

If (ϱ, u) is a classical Hölder solution to (3.6) on $[0, T]$, the pair $(\varrho, v = \Theta_\varrho^* u)$ solves then the system

$$\begin{cases} \mathcal{A}(\varrho)v = 0 & \text{in } \mathbb{D}, \ t \in [0, T], \\ v = \gamma \mathcal{K}(\varrho) - K\iota^2 \dfrac{1}{(1+\varrho)^2} - \dfrac{\omega^2}{2}(1+\varrho)^2 & \text{on } \mathbb{S}, \ t \in [0, T], \\ \partial_t \varrho = \mathcal{B}(\varrho, v) & \text{on } \mathbb{S}, \ t \in [0, T], \\ \varrho(0) = \varrho_0 & \text{on } \mathbb{S}, \end{cases}$$
(3.16)

where
$$\mathcal{K} : \mathcal{V} \subset h^{4+\alpha}(\mathbb{S}) \to h^{2+\alpha}(\mathbb{S}), \qquad \mathcal{K}(\varrho) := \kappa_{\Gamma_\varrho} \circ \Theta_\varrho,$$

is the pulled back curvature. The notion of solution to problem (3.16) is defined similarly to that of solution to (3.6). Furthermore, the systems (3.6) and (3.16) are equivalent in the following sense:

Lemma 3.3.1 *If (ϱ, u) is a classical Hölder solution of (3.6), then $(\varrho, v := \Theta_\varrho^* u)$ is a classical Hölder solution of (3.16). Conversely, if (ϱ, v) is a classical Hölder solution of (3.16), then $(\varrho, u := \Theta_*^\varrho v)$ solves (3.6).*

Proof The main difficulty lies in proving that $v \in buc^{2+\alpha}(\mathbb{D})$ if $(\varrho, u) \in \mathcal{V} \times buc^{2+\alpha}(\Omega_\varrho)$ and that $u \in buc^{2+\alpha}(\Omega_\varrho)$ if $(\varrho, v) \in \mathcal{V} \times buc^{2+\alpha}(\mathbb{D})$. To do this, fix $\varrho \in \mathcal{V}$, $u \in buc^{2+\alpha}(\Omega_\varrho)$, and choose a sequence $(\varrho_m) \subset C^\infty(\mathbb{S})$ with $\varrho_m \nearrow \varrho$ in $h^{4+\alpha}(\mathbb{S})$. There exists also a sequence $(u_n) \subset BUC^\infty(\Omega_\varrho)$ which converges to u in $BUC^{2+\alpha}(\Omega_\varrho)$. Now let $v_{n,m} := \Theta_{\varrho_m}^* u_n \in BUC^\infty(\mathbb{D})$. By the mean value theorem and (3.8) we obtain that $v := \Theta_\varrho^* u$ is the limit in $BUC^{2+\alpha}(\mathbb{D})$ of a subsequence of $(v_{n,m})$, i.e. $v \in buc^{2+\alpha}(\mathbb{D})$. Indeed, for $n, m \in \mathbb{N}$, we write

$$v_{n,m} - v = u_n \circ \Theta_{\varrho_m} - u \circ \Theta_\varrho = (u_n \circ \Theta_{\varrho_m} - u_n \circ \Theta_\varrho) + (u_n \circ \Theta_\varrho - u \circ \Theta_\varrho)$$

$$= \left(\int_0^1 \partial u_n((1-t)\Theta_{\varrho_m} + t\Theta_\varrho) \, dt \right) (\Theta_\varrho - \Theta_{\varrho_m}) + (u_n - u) \circ \Theta_\varrho.$$

The second term of the right hand side can be made small for n sufficiently large. Taking into consideration that u_n is smooth, we can bound the $BUC^{4+\alpha}(\mathbb{D})$–norm of the integral by a constant depending only on n and ϱ. By choosing m large we can get, in view of (3.8), also the first addend small. Hence, any neighbourhood of v contains functions belonging to $BUC^\infty(\mathbb{D})$. The converse follows by a similar argument.

The curvature operator $\mathcal{K}$ maps analytically $\mathcal{V}$ into $h^{2+\alpha}(\mathbb{S})$. More exactly, we state:

Lemma 3.3.2 *We have $\mathcal{K} \in C^\omega(\mathcal{V}, h^{2+\alpha}(\mathbb{S}))$. Moreover, the derivative of $\mathcal{K}$ in 0 is given by the relation*

$$\partial \mathcal{K}(0)[\varrho] = -\varrho - \varrho'' \quad \text{for} \quad \varrho \in h^{4+\alpha}(\mathbb{S}). \tag{3.17}$$

Proof We consider first the analytic mapping

$$f : \mathbb{R}^3 \to \mathbb{R}, \quad f(x) = \frac{(1+x_1)^2 + 2x_2^2 - (1+x_1)x_3}{((1+x_1)^2 + x_2^2)^{3/2}}.$$

Since $\mathcal{K}(\varrho) = f(\varrho, \varrho', \varrho'')$, we know, by [12, Example 4.3.6] that $\mathcal{K}$, as operator form $\mathcal{V}$ into $h^{2+\alpha}(\mathbb{S})$, has the same regularity as f, so that $\mathcal{K} \in C^\omega(\mathcal{V}, h^{2+\alpha}(\mathbb{S}))$. A simple calculation shows that

$$\partial \mathcal{K}(0)[\varrho] = \nabla f(0) \cdot (\varrho, \varrho', \varrho'') = (-1, 0, -1) \cdot (\varrho, \varrho', \varrho'') = -\varrho - \varrho''$$

for all $\varrho \in h^{4+\alpha}(\mathbb{S})$. This completes the proof.

Chapter 4

The evolution equation

In this chapter we transform the system (3.16) into an evolution equation on the unit circle $\mathbb{S}$. General results of the theory of maximal regularity results, due to Sinestrari [61], can be used to prove existence of a unique classical solution, corresponding to small initial data.

4.1 The evolution equation

We start by solving the quasilinear Dirichlet problem consisting of the first two equations of (3.16). Assume first that relation (2.26) is not satisfied. Then, given $\varrho \in \mathcal{V}$, the boundary curve

$$\mathcal{C}_\varrho := \left(\Gamma_\varrho, \gamma \kappa_{\Gamma_\varrho} - K\iota^2 \frac{1}{|y|^2} - \frac{\omega^2}{2}|y|^2 \right) \subset \mathbb{R}^3$$

satisfies a bounded slope condition since the curvature of Γ_ϱ is everywhere positive and the boundary Γ_ϱ, respectively the boundary value $\gamma \kappa_{\Gamma_\varrho} - K\iota^2/|y|^2 - \omega^2|y|^2/2$, belong to the class C^2. This means that for each point $P \in \mathcal{C}_\varrho$ the curve is bounded from above and below on the cylinder $\Gamma_\varrho \times \mathbb{R}$ by two planes and coincides with them at P. The slopes of the planes are uniformly bounded by a constant which does not depend on the point P.

In view of [37, Theorem 10.5] (or [49, Theorem 6.4.2]) we obtain that for each $\iota \in (0, \infty)$ and $\varrho \in \mathcal{V}$ the Dirichlet problem

$$\begin{aligned} \mathcal{Q}u &= 0 & \text{in} \quad \Omega_\rho, \\ u &= \gamma \kappa_{\Gamma_\varrho} - K\iota^2 \frac{1}{|y|^2} - \frac{\omega^2}{2}|y|^2 & \text{on} \quad \Gamma_\rho, \end{aligned} \quad (4.1)$$

possesses at least a solution $u \in BUC^{2+\alpha}(\Omega_\rho)$. Furthermore, if (2.26) holds, then $\mathcal{Q}$ is uniformly elliptic and the same result for (4.1) is obtained by applying [49, Theorem 4.8.3]. The mean value theorem (see [37, Theorem 10.2] for more details) yields that this solution is also unique. Notice also that, if $\rho \in C^\infty(\mathbb{S})$, then u, the solution to (4.1), is also smooth, that is $u \in BUC^\infty(\Omega_\rho)$.

Summarising, we obtain with Lemma 3.3.2 the following existence, uniqueness and regularity result:

Theorem 4.1.1 *Given $\iota \in (0, \infty)$ and $\varrho \in \mathcal{V}$, there exists a unique solution $\mathcal{T}(\iota, \varrho) \in buc^{2+\alpha}(\mathbb{D})$ of the quasilinear Dirichlet problem*

$$\mathcal{A}(\varrho)v = 0 \quad \text{in } \mathbb{D},$$
$$v = \gamma \mathcal{K}(\varrho) - K\iota^2 \frac{1}{(1+\varrho)^2} - \frac{\omega^2}{2}(1+\varrho)^2 \quad \text{on } \mathbb{S}. \tag{4.2}$$

Moreover, the mapping $[(0, \infty) \times \mathcal{V} \ni (\iota, \varrho) \mapsto \mathcal{T}(\iota, \varrho) \in buc^{2+\alpha}(\mathbb{D})]$ is analytic.

Proof In view of the facts discussed above it suffices to prove that $\mathcal{T}$ is analytic in $BUC^{2+\alpha}(\mathbb{D})$. Indeed, knowing that $\mathcal{T}$ is analytic, we obtain from the density of the smooth functions in $h^{4+\alpha}(\mathbb{S})$ and the inclusion $\mathcal{T}((0, \infty) \times C^\infty(\mathbb{S})) \subset BUC^\infty(\mathbb{D})$, that $\varrho \in \mathcal{V}$ ensures $\mathcal{T}(\iota, \varrho) \in buc^{2+\alpha}(\mathbb{D})$ for all $\iota \in (0, \infty)$.

We proceed and prove now that $\mathcal{T}$ depends analytically on (ι, ϱ). Let $\mathcal{S} : \mathcal{V} \times BUC^{2+\alpha}(\mathbb{D}) \to \mathcal{L}(BUC^{2+\alpha}(\mathbb{D}), BUC^\alpha(\mathbb{D}))$ be the operator defined by

$$\mathcal{S}(\varrho, v)[u] := b_{ij}(x, \varrho, \nabla v)u_{ij} + b_i(x, \varrho, \nabla v)u_i,$$

where we use the usual sum convention. Given $(\varrho, v) \in \mathcal{V} \times BUC^{2+\alpha}(\mathbb{D})$, $\mathcal{S}(\varrho, v)$ is a linear elliptic differential operator of second order. In view of relations (3.9), (3.12)-(3.14), Lemma 3.2.2 and [12, Example 4.3.6] we have that

$$\mathcal{S} \in C^\omega(\mathcal{V} \times BUC^{2+\alpha}(\mathbb{D}), \mathcal{L}(BUC^{2+\alpha}(\mathbb{D}), BUC^\alpha(\mathbb{D}))).$$

Consequently, the operator $\mathcal{I} : (0, \infty) \times \mathcal{V} \times BUC^{2+\alpha}(\mathbb{D}) \to BUC^{2+\alpha}(\mathbb{D})$, with

$$\mathcal{I}(\iota, \varrho, v) := (\mathcal{S}(\varrho, v), \text{tr})^{-1}\left(0, \gamma\mathcal{K}(\varrho) - K\iota^2\frac{1}{(1+\varrho)^2} - \frac{\omega^2}{2}(1+\varrho)^2\right),$$

depends analytically on its variables as well. We have denoted by tr the trace operator on $\mathbb{S}$. The analyticity is obtained in view of Lemma 3.3.2, by taking into consideration that the function mapping a bijective linear operator onto its inverse is analytical. Lastly, we define the nonlinear operator

$$\mathcal{F} : (0, \infty) \times \mathcal{V} \times BUC^{2+\alpha}(\mathbb{D}) \to BUC^{2+\alpha}(\mathbb{D}), \qquad \mathcal{F}(\iota, \varrho, v) := v - \mathcal{I}(\iota, \varrho, v).$$

We observe that the solution of $\mathcal{F}(\iota, \varrho, \cdot) = 0$ is exactly the solution $\mathcal{T}(\iota, \varrho)$ of (4.2). Hence, $[(\iota, \varrho) \mapsto (\iota, \varrho, \mathcal{T}(\iota, \varrho))]$ is a parametrisation of the 0-level set of $\mathcal{F}$. The derivative of $\mathcal{F}$ with respect to v is given by the relation

$$\partial_v \mathcal{F}(\iota, \varrho, v) = \operatorname{id}_{BUC^{2+\alpha}(\mathbb{D})} - \partial_v \mathcal{I}(\iota, \varrho, v),$$

where $\partial_v \mathcal{I}(\iota, \varrho, v)$ is a compact operator. This is due to the fact that $\mathcal{S}$ and $\mathcal{I}$ have natural extensions to operators on $\mathcal{V} \times BUC^{1+\alpha}(\mathbb{D})$ and $(0, \infty) \times \mathcal{V} \times BUC^{1+\alpha}(\mathbb{D})$, respectively, and the embedding $BUC^{2+\alpha}(\mathbb{D}) \hookrightarrow BUC^{1+\alpha}(\mathbb{D})$ is compact. Consequently, $\partial_v \mathcal{F}(\iota, \varrho, v)$ is a Fredholm operator of index ind $\partial_v \mathcal{F}(\iota, \varrho, v) = 0$. Let us compute the Fréchet derivative $\partial_v \mathcal{I}(\iota, \varrho, \mathcal{T}(\iota, \varrho))$. For $(\iota, \varrho) \in (0, \infty) \times \mathcal{V}$, we set $v := \mathcal{T}(\iota, \varrho)$. Given $w \in BUC^{2+\alpha}(\mathbb{D})$, the mapping $\partial_v \mathcal{I}(\iota, \varrho, v)[w]$ is the unique solution of the following Dirichlet problem

$$\begin{cases} b_{ij}(y, \varrho, \nabla v) z_{ij} + b_i(y, \varrho, \nabla v) z_i &= -\left(\overline{\partial} b_{ij}(y, \varrho, \nabla v) v_{ij} + \overline{\partial} b_i(y, \varrho, \nabla v) v_i\right) \\ & \quad (\nabla(\Theta_*^\varrho w)(\Theta_\varrho)) \quad \text{in} \quad \mathbb{D}, \\ z &= 0 \quad \text{on} \quad \mathbb{S}, \end{cases} \quad (4.3)$$

where

$$\overline{\partial} b_{ij}(y, \varrho, \nabla v) = \psi_{\varrho,k}^i(\Theta_\varrho(y)) \psi_{\varrho,l}^j(\Theta_\varrho(y)) \partial a_{kl}(\nabla(\Theta_*^\varrho v)(\Theta_\varrho(y))), \ 1 \leq i, j \leq 2,$$

$$\overline{\partial} b_i(y, \varrho, \nabla v) = \psi_{\varrho,kl}^i(\Theta_\varrho(y)) \partial a_{kl}(\nabla(\Theta_*^\varrho v)(\Theta_\varrho(y))), \ 1 \leq i \leq 2,$$

and ∂a_{ij} are the usual Fréchet derivatives of the real analytic functions $a_{ij} : \mathbb{R}^2 \to \mathbb{R}$, $1 \leq i, j \leq 2$. We state that the Fréchet derivative $\partial_v \mathcal{F}(\iota, \varrho, \mathcal{T}(\iota, \varrho))$ is an isomorphism for all $(\iota, \varrho) \in (0, \infty) \times \mathcal{V}$. Taking into consideration that $\partial_v \mathcal{F}(\iota, \varrho, \mathcal{T}(\iota, \varrho))$ is a Fredholm operator of index 0, it suffices to prove that it is one-to-one. Indeed, let $w \in BUC^{2+\alpha}(\mathbb{D})$ be a function with the property that $\partial_v \mathcal{F}(\iota, \varrho, \mathcal{T}(\iota, \varrho))[w] = 0$, that is $w = \partial_v \mathcal{I}(\iota, \varrho, \mathcal{T}(\iota, \varrho))[w]$. Then, w is the solution of (4.3), hence $w = 0$.

The implicit function theorem states then, that $\mathcal{T}$ is analytic in a neighbourhood of (ι, ϱ) for all $(\iota, \varrho) \in (0, \infty) \times \mathcal{V}$. This completes the proof. $\square$

With these preparations, we reduce now our problem to an evolution equation on the unit circle $\mathbb{S}$. Namely, for $(\iota, \varrho) \in (0, \infty) \times \mathcal{V}$, we insert $\mathcal{T}(\iota, \varrho)$ in the third equation of (3.16) and get the following problem

$$\partial_t \varrho = \Phi(\iota, \varrho), \qquad \varrho(0) = \varrho_0, \qquad (4.4)$$

where the nonlinear and nonlocal operator Φ has order three and is defined by

$$\Phi(\iota, \varrho) := \mathcal{B}(\varrho, \mathcal{T}(\iota, \varrho)) \quad \text{for} \quad (\iota, \varrho) \in (0, \infty) \times \mathcal{V}. \qquad (4.5)$$

Theorem 4.1.2 *The operator Φ is analytic $\Phi \in C^\omega((0, \infty) \times \mathcal{V}, h^{1+\alpha}(\mathbb{S}))$. Given $\iota \in (0, \infty)$ and $\varrho \in \mathcal{V}$, the Fréchet derivative*

$$\partial_\varrho \Phi(\iota, 0)[\varrho] = -\frac{1}{\overline{\mu}(0)} \partial_\nu \left(\partial_\varrho \mathcal{T}(\iota, 0)[\varrho]\right), \quad \varrho \in h^{4+\alpha}(\mathbb{S})), \qquad (4.6)$$

where $\nu := \nu_0$ is the outward unit normal at $\mathbb{S} = \Gamma_0$.

Proof Since $\overline{\mu} \in C^\omega((0, \infty))$, we obtain from relation (3.15) that $\mathcal{B} \in C^\omega(\mathcal{V} \times buc^{2+\alpha}(\mathbb{D}), h^{1+\alpha}(\mathbb{S}))$. The analyticity of Φ follows then together with Theorem 4.1.1. By the chain rule

$$\partial_\varrho \Phi(\iota, 0)[\varrho] = \partial \mathcal{B}(0, \mathcal{T}(0))[\varrho, \partial_\varrho \mathcal{T}(\iota, 0)[\varrho]], \qquad \varrho \in h^{4+\alpha}(\mathbb{S}).$$

We prove first that

$$\partial \mathcal{B}(0, \mathcal{T}(0))[\varrho, v] = -\frac{1}{\overline{\mu}(0)} \partial_\nu v \quad \text{for all } (\varrho, v) \in h^{4+\alpha}(\mathbb{S}) \times buc^{2+\alpha}(\mathbb{D}). \qquad (4.7)$$

Having stated (4.7), relation (4.6) is then immediate. We proceed with the proof of (4.7). Given $(\varrho, v) \in \mathcal{V} \times buc^{2+\alpha}(\mathbb{D})$, we obtain from (3.4) and (3.15) that

$$\langle \nabla(\Theta^\varrho_* v), \nabla N_\varrho \rangle(\Theta_\varrho) = \left[x_i \psi^j_{\varrho,i}(\Theta_\varrho) + \frac{\varrho'}{1+\varrho} \left(x_2 \psi^j_{\varrho,1}(\Theta_\varrho) - x_1 \psi^j_{\varrho,2}(\Theta_\varrho) \right) \right] v_j, \qquad (4.8)$$

where $x = (x_1, x_2) \in \mathbb{S}$. The solution of the Dirichlet problem (4.2) is the constant $\mathcal{T}(\iota, 0) = \gamma - K\iota^2 - \omega^2/2$, so that $\mathcal{B}(0, \mathcal{T}(\iota, 0)) = 0$. Furthermore, given $(\varrho, v) \in h^{4+\alpha}(\mathbb{S}) \times \textit{buc}^{2+\alpha}(\mathbb{D})$, it is convenient to write

$$\mathcal{B}(\varrho, \mathcal{T}(\iota, 0) + v) - \mathcal{B}(0, \mathcal{T}(\iota, 0)) + \frac{1}{\overline{\mu}(0)} \langle \nabla(\Theta_*^0 v), \nabla N_0 \rangle (\Theta_0) =$$

$$= -\left(\frac{1}{\overline{\mu}(|\nabla(\Theta_*^\varrho v)|^2)} - \frac{1}{\overline{\mu}(0)} \right) \langle \nabla(\Theta_*^\varrho v), \nabla N_\varrho \rangle (\Theta_\varrho)$$

$$- \frac{1}{\overline{\mu}(0)} \left(\langle \nabla(\Theta_*^\varrho v), \nabla N_\varrho \rangle (\Theta_\varrho) - \langle \nabla(\Theta_*^0 v), \nabla N_0 \rangle (\Theta_0) \right).$$

By the mean value theorem and (4.8) we obtain that the first addend form the right hand side of this identity is of order $O(\|(\varrho, v)\|^2_{C^{4+\alpha}(\mathbb{S}) \times BUC^{2+\alpha}(\mathbb{D})})$ for (ϱ, v) small enough. For the second term of the sum from the right hand side of the preceding identity we obtain, in view of the same relation (4.8), that

$$\langle \nabla(\Theta_*^\varrho v), \nabla N_\varrho \rangle (\Theta_\varrho) - \langle \nabla(\Theta_*^0 v), \nabla N_0 \rangle (\Theta_0) =$$

$$= \left\{ x_i (\psi_{\varrho,i}^j(\Theta_\varrho) - \psi_{0,i}^j(\Theta_0)) + \frac{\varrho'}{1+\varrho} \left(x_2 \psi_{\varrho,1}^j(\Theta_\varrho) - x_1 \psi_{\varrho,2}^j(\Theta_\varrho) \right) \right\} v_j,$$

so that the difference $\mathcal{B}(\varrho, \mathcal{T}(\iota, 0) + v) - (-1/\overline{\mu}(0) \langle \nabla(\Theta_*^0 v), \nabla N_0 \rangle (\Theta_0))$ is of order $O(\|(\varrho, v)\|^2_{C^{4+\alpha}(\mathbb{S}) \times BUC^{2+\alpha}(\mathbb{D})})$ for small (ϱ, v). The desired formula (4.7) follows in virtue of $\Theta_0 = \text{id}_{\mathbb{R}^2}$ and $\nabla N_0 = \nu$, the unit outward normal at $\mathbb{S}$. $\square$

For later purposes, we need to consider more closely the formula (4.6). Therefore, we determine now the Fréchet derivative of $\mathcal{T}$ with respect to ϱ at $(\iota, 0)$, $\iota \in (0, \infty)$.

Lemma 4.1.3 *Given $\iota \in (0, \infty)$ and $\varrho \in h^{4+\alpha}(\mathbb{S})$, $\partial_\varrho \mathcal{T}(\iota, 0)[\varrho] \in \textit{buc}^{2+\alpha}(\mathbb{D})$ is the unique solution of the Dirichlet problem*

$$\begin{cases} \Delta w = 0 & \text{in } \mathbb{D}, \\ w = (2K\iota^2 - \gamma - \omega^2)\varrho - \gamma \varrho'' & \text{on } \mathbb{S}. \end{cases} \quad (4.9)$$

Proof Setting $v := \mathcal{T}(\iota, \varrho)$ and $z := \mathcal{T}(\iota, \varrho) - \mathcal{T}(\iota, 0) - w$, we obtain, in view of $\mathcal{T}(\iota, 0) = \gamma - K\iota^2 - \omega^2/2$ and $b_{ij}(y, 0, 0) = \delta_{ij}/\overline{\mu}(0)$, $b_i(y, 0, 0) = 0$, that

$$0 = \mathcal{A}(\varrho)\mathcal{T}(\iota, \varrho) - \mathcal{A}(0)\mathcal{T}(\iota, 0) - \frac{1}{\overline{\mu}(0)}\Delta w =$$
$$= b_{ij}(y, \varrho, \nabla v)v_{ij} + b_i(y, \varrho, \nabla v)v_i - b_{ij}(y, 0, 0)w_{ij} - b_i(y, 0, 0)w_i$$
$$= b_{ij}(y, \varrho, \nabla v)z_{ij} + b_i(y, \varrho, \nabla v)z_i$$
$$\quad + (b_{ij}(y, \varrho, \nabla v) - b_{ij}(y, 0, 0))w_{ij} + (b_i(y, \varrho, \nabla v) - b_i(y, 0, 0))w_i.$$

Hence, z is solution of the equation

$$b_{ij}(y, \varrho, \nabla v)z_{ij} + b_i(y, \varrho, \nabla v)z_i = -(b_{ij}(y, \varrho, \nabla v) - b_{ij}(y, 0, 0))w_{ij}$$
$$- (b_i(y, \varrho, \nabla v) - b_i(y, 0, 0))w_2 \text{ in } \mathbb{D}.$$

Moreover, from Lemma 3.3.2 and the relations

$$\partial\left(\frac{1}{(1+\varrho)^2}\right)\Big|_{\varrho=0}[\varrho] = -2\varrho, \quad \partial\left((1+\varrho)^2\right)\Big|_{\varrho=0}[\varrho] = 2\varrho, \quad \varrho \in h^{4+\alpha}(\mathbb{S}),$$

the Fréchet derivative with respect to ϱ of the boundary value of the Dirichlet problem (4.2) is exactly the boundary value of (4.9), i.e.

$$\partial_\varrho\left(\gamma\mathcal{K}(\varrho) - K\iota^2\frac{1}{(1+\varrho)^2} - \frac{\omega^2}{2}(1+\varrho)^2\right)\Big|_{\varrho=0}[\varrho] = (2K\iota^2 - \gamma - \omega^2)\varrho - \gamma\varrho''$$

for all $\varrho \in h^{4+\alpha}(\mathbb{S})$, so that $\|\operatorname{tr} z\|_{C^{2+\alpha}(\mathbb{S})} = O(\|\varrho\|_{C^{4+\alpha}(\mathbb{S})}^2)$ for small $\varrho \in h^{4+\alpha}(\mathbb{S})$.

Using estimates for solution of elliptic problems, cf. [37, Theorem 3.7 and Theorem 6.6], we proved that

$$\mathcal{T}(\iota, \varrho) - \mathcal{T}(\iota, 0) - w = O(\|\varrho\|_{C^{4+\alpha}(\mathbb{S})}^2) \quad \text{for small} \quad \varrho \in h^{4+\alpha}(\mathbb{S}),$$

and we are done. $\square$

It is worth noticing that, given $\varrho \in h^{4+\alpha}(\mathbb{S})$, the solution $\partial_\varrho\mathcal{T}(\iota, 0)[\varrho]$ to (4.9) can be explicitly determined by Poisson's formula. By combining Theorem 4.1.2 and Lemma 4.1.3 we find therefore, that $\partial_\varrho\Phi(\iota, 0)$ is a

Fourier multiplication operator. To this scope we consider Fourier expansions $\varrho = \sum_{k\in\mathbb{Z}} \widehat{\varrho}(k) x^k$ of the functions $\varrho \in h^{4+\alpha}(\mathbb{S})$, where the k–th Fourier coefficient $\widehat{\varrho}(k)$ of ϱ is defined by

$$\widehat{\varrho}(k) = \int_{\mathbb{S}} \varrho(x) x^{-k}\, dx = \frac{1}{2\pi} \int_0^{2\pi} \varrho(t) e^{-ikt}\, dt, \quad k \in \mathbb{Z}.$$

It is known that for continuously differentiable functions the series converges in $C(\mathbb{S})$, so that, when we write $\varrho = \sum_{k\in\mathbb{Z}} \widehat{\varrho}(k) x^k \in h^{4+\alpha}(\mathbb{S})$, we mean that $\varrho \in h^{4+\alpha}(\mathbb{S})$ has the the Fourier expansion $\varrho = \sum_{k\in\mathbb{Z}} \widehat{\varrho}(k) x^k$, which converges to ϱ at least in $C^3(\mathbb{S})$.

Corollary 4.1.4 *Given $\iota \in (0, \infty)$ and $k \in \mathbb{Z}$, let*

$$\lambda_k(\iota) := -\frac{1}{\overline{\mu}(0)} |k| \left((2K\iota^2 - \gamma - \omega^2) + \gamma k^2 \right). \tag{4.10}$$

Then $\partial_\varrho \Phi(\iota, 0)$ is a Fourier multiplication operator with symbol $(\lambda_k(\iota))_{k\in\mathbb{Z}}$ given by (4.10), i.e.

$$\partial_\varrho \Phi(\iota, 0) \left[\sum_{k\in\mathbb{Z}} \widehat{\varrho}(k) x^k \right] = \sum_{k\in\mathbb{Z}} \lambda_k(\iota) \widehat{\varrho}(k) x^k \quad \text{for } \varrho = \sum_{k\in\mathbb{Z}} \widehat{\varrho}(k) x^k \in h^{4+\alpha}(\mathbb{S}). \tag{4.11}$$

Proof Let $\varrho = \sum_{k\in\mathbb{Z}} \widehat{\varrho}(k) x^k$ be given. Then

$$(2K\iota^2 - \gamma - \omega^2)\varrho - \gamma \varrho'' = \sum_{k\in\mathbb{Z}} \left[(2K\iota^2 - \gamma - \omega^2) + \gamma k^2 \right] \widehat{\varrho}(k) x^k \quad \text{on } \mathbb{S},$$

and from Poisson's formula (see [41, relation (7.28)]) we get

$$\partial_\varrho (\mathcal{T}(\iota, 0))[\varrho](rx) = \sum_{k\in\mathbb{Z}} \left[(2K\iota^2 - \gamma - \omega^2) + \gamma k^2 \right] r^{|k|} x^k$$

for all $r \in [0, 1]$ and $x \in \mathbb{S}$. The desired conclusion follows then from (4.6). □

4.2 The well-posedness result

In this section we fix $0 < \beta < \alpha$, $\iota, \omega, \gamma \in (0, \infty)$ and prove first that the operator $\partial_\varrho \Phi(\iota, 0)$, considered as an unbounded operator in $h^{1+\beta}(\mathbb{S})$ with domain $h^{4+\beta}(\mathbb{S})$, generates a strongly continuous and analytic semigroup in $\mathcal{L}(h^{1+\beta}(\mathbb{S}))$, i.e. $-\partial_\varrho \Phi(\iota, 0) \in \mathcal{H}(h^{4+\beta}(\mathbb{S}), h^{1+\beta}(\mathbb{S}))$ (see [4]). As a matter of fact, we know that $\partial_\varrho \Phi(\iota, 0) \in \mathcal{L}(h^{4+\beta}(\mathbb{S}), h^{1+\beta}(\mathbb{S}))$ since α, the constant fixed from the start, was arbitrarily picked in $(0, 1)$. We state:

Theorem 4.2.1 *Given* $0 < \beta < \alpha$, $\iota \in (0, \infty)$, *we have that*

$$-\partial_\varrho \Phi(\iota, 0) \in \mathcal{H}(h^{4+\beta}(\mathbb{S}), h^{1+\beta}(\mathbb{S})).$$

Having established this result, the proof of Theorem 3.2.3 follows more or less directly from the local existence, uniqueness and regularity results on parabolic problems presented in [51]. However, the proof of Theorem 4.2.1 is rather involved and needs some preliminary results.

In the following we consider the complex Banach spaces $h^{m+\beta}(\mathbb{S}, \mathbb{C})$, $m = 1, 4$ and show that the complexification of $\partial_\varrho \Phi(\iota, 0)$, which we denote by A, satisfies $-A \in \mathcal{H}(h^{4+\alpha}(\mathbb{S}, \mathbb{C}), h^{1+\alpha}(\mathbb{S}, \mathbb{C}))$.

Using the same notations as in [4], we have $h^{4+\beta}(\mathbb{S}, \mathbb{C}) \stackrel{d}{\hookrightarrow} h^{1+\beta}(\mathbb{S}, \mathbb{C})$ and, given $\mathfrak{k} \geq 1$ and $\mathfrak{w} > 0$, we write

$$-A \in \mathcal{H}(h^{4+\beta}(\mathbb{S}, \mathbb{C}), h^{1+\beta}(\mathbb{S}, \mathbb{C}), \mathfrak{k}, \mathfrak{w})$$

if $\mathfrak{w} - A \in \mathcal{L}is(h^{4+\beta}(\mathbb{S}, \mathbb{C}), h^{1+\beta}(\mathbb{S}, \mathbb{C}))$ and

$$\mathfrak{k}^{-1} \leq \frac{\|(\lambda - A)[\varrho]\|_{C^{1+\beta}(\mathbb{S},\mathbb{C})}}{|\lambda| \|\varrho\|_{C^{1+\beta}(\mathbb{S},\mathbb{C})} + \|\varrho\|_{C^{4+\beta}(\mathbb{S})}} \leq \mathfrak{k}, \quad \varrho \in h^{4+\beta}(\mathbb{S}, \mathbb{C}) \setminus \{0\}, \quad \operatorname{Re} \lambda \geq \mathfrak{w}.$$

By [4, Theorem 1.2.2]

$$\mathcal{H}(h^{4+\beta}(\mathbb{S}, \mathbb{C}), h^{1+\beta}(\mathbb{S}, \mathbb{C})) = \bigcup_{\substack{\mathfrak{k} \geq 1 \\ \mathfrak{w} > 0}} \mathcal{H}(h^{4+\beta}(\mathbb{S}, \mathbb{C}), h^{1+\beta}(\mathbb{S}, \mathbb{C}), \mathfrak{k}, \mathfrak{w}),$$

so that it suffices to prove that $-A \in \mathcal{H}(h^{4+\beta}(\mathbb{S}, \mathbb{C}), h^{1+\beta}(\mathbb{S}, \mathbb{C})), \mathfrak{k}, \mathfrak{w})$, for some $\mathfrak{k} \geq 1$ and $\mathfrak{w} > 0$. In view of [4, Remark 1.2.1 (a)], we are done if we find $\mathfrak{k} \geq 1$ and $\mathfrak{w} > 0$ such that

$$\lambda - A \in \mathcal{L}is(h^{4+\beta}(\mathbb{S}, \mathbb{C}), h^{1+\beta}(\mathbb{S}, \mathbb{C})), \tag{4.12}$$

$$|\lambda| \cdot \|R(\lambda, A)\|_{\mathcal{L}(h^{1+\beta}(\mathbb{S},\mathbb{C}))} \leq \mathfrak{k} \tag{4.13}$$

for all $\operatorname{Re} \lambda \geq \mathfrak{w}$. Given λ in the resolvent set $\rho(A)$ of A, we denote by $R(\lambda, A)$ the resolvent operator of A. Fix

$$\mathfrak{w} := \max\left(\{4\lambda_k : k \in \mathbb{Z}\} \cup \{4\}\right). \tag{4.14}$$

Lemma 4.2.2 *Presuppose* $R(\lambda, A) \in \mathcal{L}(C^{1+\beta}(\mathbb{S}, \mathbb{C}), C^{m+\beta}(\mathbb{S}, \mathbb{C}))$ *for* $\operatorname{Re} \lambda \geq \mathfrak{w}$ *and* $m \in \{1, 4\}$. *Then* $R(\lambda, A) \in \mathcal{L}(h^{1+\beta}(\mathbb{S}, \mathbb{C}), h^{m+\beta}(\mathbb{S}, \mathbb{C}))$.

Proof From the assumption we get $R(\lambda, A) \in \mathcal{L}(h^{1+\beta}(\mathbb{S}, \mathbb{C}), C^{m+\beta}(\mathbb{S}, \mathbb{C}))$. Moreover, it is clear from (4.11) that the resolvent $R(\lambda, A)$ is the Fourier multiplication operator

$$\sum_{k \in \mathbb{Z}} \widehat{\varrho}(k) x^k \mapsto \sum_{k \in \mathbb{Z}} \frac{1}{\lambda - \lambda_k(\iota)} \widehat{\varrho}(k) x^k.$$

The constant $\mathfrak{w}$ has been chosen such that $|\lambda - \lambda_k(\iota)| \geq 3$ for all $k \in \mathbb{Z}$ and $\operatorname{Re} \lambda \geq \mathfrak{w}$. Whence, if $\varrho \in C^\infty(\mathbb{S}, \mathbb{C})$, it folows from relation (2.34) that $R(\lambda, A)\varrho \in C^\infty(\mathbb{S}, \mathbb{C})$. The assertion follows due to the density of $C^\infty(\mathbb{S}, \mathbb{C})$ in $h^{m+\beta}(\mathbb{S}, \mathbb{C})$.
$\square$

Proof (Proof of Theorem 4.2.1) We begin by verifying that the relations (4.12) and (4.13) are valid if $\mathfrak{w}$ is defined by (4.14). We start with (4.12). Due to Lemma 4.2.2, we show only $R(\lambda, A) \in \mathcal{L}(C^{1+\beta}(\mathbb{S}, \mathbb{C}), C^{4+\beta}(\mathbb{S}, \mathbb{C}))$. Recall that, given $\varrho \in C^{1+\beta}(\mathbb{S}, \mathbb{C})$, the resolvent

$$R(\lambda, A)\varrho = \sum_{k \in \mathbb{Z}} M_k^\lambda \widehat{\varrho}(k) x^k, \quad \text{where} \quad M_k^\lambda := \frac{1}{\lambda - \lambda_k(\iota)},$$

so that (4.12) is true if the sequence $(M_k^\lambda)_k$ satisfies the assumptions $(i) - (iii)$ of Theorem 2.2.1 with $s = 1 + \beta$ and $r = 4 + \beta$. Condition (i) follows in virtue of

$$\lim_{k \to \infty} \frac{k^3}{\lambda_k(\iota)} = -\frac{\overline{\mu}(0)}{\gamma}.$$

In view of $|\lambda_{k+1}(\iota) - \lambda_k(\iota)|/k^2 \to 3\gamma/\overline{\mu}(0)$, we further have

$$\lim_{k\to\infty} k^4 \left| \frac{1}{\lambda - \lambda_{k+1}(\iota)} - \frac{1}{\lambda - \lambda_k(\iota)} \right|$$

$$= \lim_{k\to\infty} \frac{|k|^3}{|\lambda_{k+1}(\iota)|} \frac{|k|^3}{|\lambda_k(\iota)|} \frac{|\lambda_{k+1}(\iota) - \lambda_k(\iota)|}{k^2} = \frac{3\overline{\mu}(0)}{\gamma},$$

and (ii) is also satisfied. Finally, in virtue of

$$\lim_{k\to\infty} k^5 \left| \frac{1}{\lambda - \lambda_{k+2}(\iota)} - \frac{2}{\lambda - \lambda_{k+1}(\iota)} + \frac{1}{\lambda - \lambda_k(\iota)} \right| = \frac{12\overline{\mu}(0)}{\gamma}$$

we conclude that all the assumptions of Theorem 2.2.1 are satisfied, hence $\{\lambda : \operatorname{Re}\lambda \geq \mathfrak{w}\} \subset \rho(A)$.

In order to prove (4.13), we have to check that the family $\{|\lambda|R(\lambda, A) : \operatorname{Re}\lambda \geq \mathfrak{w}\}$, with

$$|\lambda|R(\lambda, A)\varrho = \sum_{k\in\mathbb{Z}} N_k^\lambda \widehat{\varrho}(k) x^k, \quad \text{and} \quad N_k^\lambda := \frac{|\lambda|}{\lambda - \lambda_k(\iota)},$$

is uniformly bounded in $\mathcal{L}(C^{1+\beta}(\mathbb{S}, \mathbb{C}))$. If $\operatorname{Re}\lambda \geq \omega$, then from the definition of $\mathfrak{w}$ we get $2|\lambda - \lambda_k(\iota)|^2 \geq |\lambda|^2$ for all $k \in \mathbb{Z}$, hence

$$\sup_{\operatorname{Re}\lambda \geq \mathfrak{w}} \sup_{k\in\mathbb{Z}} |N_k^\lambda| \leq \sqrt{2}.$$

We also have that $|\lambda - \lambda_k(\iota)| \geq |\lambda_k(\iota)|$ if $\operatorname{Re}\lambda \geq \mathfrak{w}$ and $k \in \mathbb{Z}$. Thus

$$|k||N_{k+1}^\lambda - N_k^\lambda| = \frac{|\lambda|}{|\lambda - \lambda_{k+1}(\iota)|} \frac{|k|^3}{|\lambda - \lambda_k(\iota)|} \frac{|\lambda_{k+1}(\iota) - \lambda_k(\iota)|}{k^2}$$

$$\leq \sqrt{2} \frac{|\lambda_{k+1}(\iota) - \lambda_k(\iota)|}{k^2} x_k,$$

where

$$x_k := \begin{cases} \dfrac{|k|^3}{|\lambda_k(\iota)|}, & \lambda_k(\iota) \neq 0, \\ |k|^3, & \lambda_k(\iota) = 0. \end{cases}$$

Moreover, the set $\{k \in \mathbb{Z} : \lambda_k(\iota) = 0\}$ is bounded, $|k|^3/|\lambda_k(\iota)| \to_{|k|\to\infty} \overline{\mu}(0)/\gamma$, hence

$$\sup_{\operatorname{Re}\lambda \geq \mathfrak{w}} \sup_{k\in\mathbb{Z}} |k||N_{k+1}^\lambda - N_k^\lambda| < \infty.$$

Finally,
$$k^2|N_{k+2}^\lambda - 2N_{k+1}^\lambda + N_k^\lambda| \leq 2x_k \frac{|\lambda_{k+2}(\iota) - 2\lambda_{k+1}(\iota) + \lambda_k(\iota)|}{|k|} +$$
$$+ \sqrt{2}x_k x_{k+1} \left| \frac{\lambda_k(\iota)(\lambda_{k+1}(\iota) - \lambda_{k+2}(\iota))}{k^4} \right.$$
$$\left. + \frac{\lambda_{k+2}(\iota)(\lambda_{k+1}(\iota) - \lambda_k(\iota))}{k^4} \right|,$$

and taking into consideration $(\lambda_{k+2}(\iota) - 2\lambda_{k+1}(\iota) + \lambda_k(\iota))/k \xrightarrow[|k|\to\infty]{} -6\gamma/\overline{\mu}(0)$, and
$$\frac{|\lambda_k(\iota)(\lambda_{k+1}(\iota) - \lambda_{k+2}(\iota)) + \lambda_{k+2}(\iota)(\lambda_{k+1}(\iota) - \lambda_k(\iota))|}{k^4} \to_{|k|\to\infty} 12\gamma^2/\overline{\mu}(0)^2,$$

we obtain (4.13).

We have shown that $-A \in \mathcal{H}(h^{4+\beta}(\mathbb{S},\mathbb{C}), h^{1+\beta}(\mathbb{S},\mathbb{C})), \mathfrak{k}, \mathfrak{w})$ for some $\mathfrak{k} > 1$. Denote by $\{e^{tA}\}_{t\geq 0} \subset \mathcal{L}(h^{1+\beta}(\mathbb{S},\mathbb{C}))$ the strongly continuous and analytic semigroup generated by A. In virtue of [51, Corollary 2.1.3], the restriction $\{e^{tA}|_{h^{1+\beta}(\mathbb{S})}\}_{t\geq 0}$, is a strongly continuous and analytic semigroup in $\mathcal{L}(h^{1+\beta}(\mathbb{S}))$ having $\partial_\varrho \Phi(\iota, 0)$ as generator. □

With these preparations done, we come now to the proof of Theorem 3.2.3:

Proof (Proof of Theorem 3.2.3) Let $\beta < \alpha$ be fixed. We infer from Theorem 4.2.1 that the derivative $\partial_\varrho \Phi(\iota, 0)$ generates a strongly continuous and analytic semigroup in $\mathcal{L}(h^{1+\beta}(\mathbb{S}))$. The set of generators of such semigroups is open in $\mathcal{L}(h^{1+\beta}(\mathbb{S}))$ (cf. [4, Theorem 1.3.1]), so that, since Φ is analytic, we find an open neighbourhood $\widetilde{\mathcal{O}}$ of the zero function in $h^{4+\beta}(\mathbb{S})$ with the property that $-\partial_\varrho \Phi(\iota, \varrho) \in \mathcal{H}(h^{4+\beta}(\mathbb{S}), h^{1+\beta}(\mathbb{S}))$ for all $\varrho \in \widetilde{\mathcal{O}}$.

The interpolation property (2.42) of the small Hölder spaces implies that the derivative $\partial_\varrho \Phi(\iota, \varrho)$ is for each $\varrho \in \mathcal{O} := \widetilde{\mathcal{O}} \cap \mathcal{V} \subset h^{4+\alpha}(\mathbb{S})$, the part in $h^{1+\alpha}(\mathbb{S})$ of the operator $\partial_\varrho \Phi(\iota, \varrho) : h^{4+\beta}(\mathbb{S}) \to h^{1+\beta}(\mathbb{S})$, and belongs therfore to $-\partial_\varrho \Phi(\iota, \varrho) \in \mathcal{H}(h^{4+\alpha}(\mathbb{S}), h^{1+\alpha}(\mathbb{S}))$. Theorem 3.2.3 follows now directly from [51, Theorem 8.3.9 and Theorem 8.4.1]. □

Chapter 5

Stability properties and bifurcation results

As we noticed in Observation 3.2.4, the volume of fluid is preserved by the flow. That is why we steadily assume through out this chapter that the cell contains a volume of fluid equal to that of the unitary disc.

We are interested in determining the equilibria of the problem (3.6) and to study their stability properties. We prove first that the trivial equilibrium, when the fluid domain is the unitary discus, is exponentially stable when the current intensity ι exceeds a certain value ι_*. For $\iota < \iota_*$, the trivial solution is unstable and we ask our selfs if this fact does not reflect in the existence of nontrivial equilibria of (3.6) arbitrarily close to this trivial equilibrium. Using ι as a bifurcation parameter, we find a finite number of global bifurcation branches consisting of stationary solutions of (3.6) showing the typical characteristics of fingering. More precisely, when the current's intensity passes certain critical values $\bar{\iota}_l$, the trivial circular equilibrium perturbs and fingers, which evolve together with the current's intensity, are formed.

5.1 Stability of the trivial equilibrium

Let us begin by considering the free boundary problem corresponding to the stationary solutions of system (3.6). Namely, if (ϱ, u) is a stationary solution of (3.6), i.e. ϱ and u do not depend on the time variable t, then

(ϱ, u) solves the following system

$$\begin{cases} \mathcal{Q}u = 0 & \text{in } \Omega_\varrho, \\ u = \gamma \kappa_{\Gamma_\varrho} - K\iota^2 \dfrac{1}{|y|^2} - \dfrac{\omega^2}{2}|y|^2 & \text{on } \Gamma_\varrho, \\ \partial_\nu u = 0 & \text{on } \Gamma_\varrho. \end{cases} \quad (5.1)$$

We notice first that the stationary solutions of (5.1) are a priori smooth.

Observation 5.1.1 *If $\varrho \in \mathcal{V}$ is a stationary solution of (5.1), then $\varrho \in C^\infty(\mathbb{S})$.*

Proof By Stokes' theorem, we obtain from the first and third equation of the system that the potential u must be a constant function. The second equation of (5.1) yields

$$\varrho'' = 1 + \varrho + \frac{2\varrho'^2}{1+\varrho} - \frac{\left(u + \dfrac{K\iota^2}{(1+\varrho)^2} + \dfrac{\omega^2}{2}(1+\varrho)^2\right)((1+\varrho)^2 + \varrho'^2)^{3/2}}{1+\varrho}, \quad (5.2)$$

and, using an induction argument, we conclude that $\varrho \in C^\infty(\mathbb{S})$. $\square$

As we mentioned in the introduction of this chapter, we look for steady states of problem (3.6) corresponding to a fluid volume equal to that of the unitary disc $\mathbb{D}$, i.e. $\text{vol}\,\Omega_\varrho = \pi$. We can describe this condition in dependence of the function ϱ parameterising the boundary Γ_ϱ. By Stokes' theorem and relation (3.4) we have

$$\text{vol}\,\Omega_\varrho = \int_{\Omega_\varrho} dy = \frac{1}{2}\int_{\Gamma_\varrho} \langle y|\nu_\varrho\rangle\, ds = \frac{1}{2}\int_{\Gamma_\varrho}(1+\varrho)\langle y|\nabla N_\varrho\rangle \frac{1}{\sqrt{(1+\varrho)^2+\varrho'^2}}\, ds$$

$$= \pi \int_{\mathbb{S}}(1+\varrho)\langle y|\nabla N_\varrho\rangle(\Theta_\varrho(x))\, dx = \pi \int_{\mathbb{S}}(1+\varrho)^2\, dx,$$

so that we look for stationary solutions of (3.6) which verify the additional condition

$$\int_{\mathbb{S}} 2\varrho + \varrho^2\, dx = 0. \quad (5.3)$$

In particular, we re-discover that there is only one circular stationary solution of (3.6). The value of the constant potential is $u = \gamma - K\iota^2 - \omega^2/2$. In the following we study the behaviour in time of the solution of the moving boundary problem (3.6), corresponding to an initial data $\varrho_0 \in \mathcal{O}$ close to 0, in dependence of the current intensity ι. However, from the Observation 3.2.4 we know that the volume of the fluid domain is preserved. This is the reason why it is natural to ask our self only what happens with the fluid domain when $\varrho_0 \in \mathcal{O}$ is not only close to 0 but also satisfies the relation (5.3).

To this scope we consider an equivalent formulation of our problem (3.6), which enables us to focus only on the behaviour of solutions to (3.6) satisfying (5.3). Consider therefore, a solution $\varrho : [0, T] \to \mathcal{V}$ of (3.6). Then

$$\partial_t \varrho = \Phi(\iota, \varrho), \quad \text{so that} \quad \partial_t \left(\frac{(1+\varrho)^2}{2} \right) = (1+\varrho)\Phi(\iota, \varrho).$$

Notice that $\int_{\mathbb{S}} (1 + \varrho)\Phi(\iota, \varrho) \, dx = 0$. Indeed, given $\varrho \in \mathcal{V}$, we set $u := \Theta_*^\varrho \mathcal{T}(\iota, \varrho)$, and obtain from equation (3.4) and Stokes' theorem, as in the proof of Observation 3.2.4, that

$$\int_{\mathbb{S}} (1+\varrho)\Phi(\iota, \varrho) \, dx = \int_{\mathbb{S}} (1+\varrho)\mathcal{B}(\varrho, \mathcal{T}(\iota, \varrho)) \, dx$$

$$= -\int_{\mathbb{S}} (1+\varrho) \left\langle \frac{\nabla u}{\bar{\mu}(|\nabla u|^2)}, \nabla N_\varrho \right\rangle (\Theta_\varrho) \, dx$$

$$= -\int_{\mathbb{S}} \left\langle \frac{\nabla u}{\bar{\mu}(|\nabla u|^2)}, \frac{\nabla N_\varrho}{|\nabla N_\varrho|} \right\rangle (\Theta_\varrho) \sqrt{(1+\varrho)^2 + \varrho'^2} \, dx$$

$$= -\frac{1}{2\pi} \int_{\Gamma_\varrho} \left\langle \frac{\nabla u}{\bar{\mu}(|\nabla u|^2)}, \nu_\varrho \right\rangle d\sigma = -\frac{1}{2\pi} \int_{\Omega_\varrho} \mathcal{Q}u \, dy = 0.$$

We define $\zeta(t) := (2\varrho(t) + \varrho^2(t))/2 \in h^{4+\alpha}(\mathbb{S})$, and find that ζ solves the equation

$$\partial_t \zeta = \Psi(\iota, \zeta), \tag{5.4}$$

where $\Psi : (0, \infty) \times \mathcal{V}_0 \subset (0, \infty) \times h_0^{4+\alpha}(\mathbb{S}) \to h_0^{1+\alpha}(\mathbb{S})$ is given by

$$\Psi(\iota, \zeta) = \sqrt{1 + 2\zeta}\, \Phi(\iota, \sqrt{1 + 2\zeta} - 1).$$

The spaces

$$h_0^{m+\beta}(\mathbb{S}) := \left\{ \zeta \in h^{m+\beta}(\mathbb{S}) : \int_{\mathbb{S}} \zeta \, dx = 0 \right\},$$

$m \in \mathbb{N}$ and $\beta \in (0,1)$, are closed subspaces of the small Hölder spaces, and $\mathcal{V}_0$ is an appropriate neighbourhood of 0 in $h_0^{4+\alpha}(\mathbb{S})$. Functions $\zeta \in h_0^{m+\beta}(\mathbb{S})$, $m \in \mathbb{N}$ and $\beta \in (0,1)$, have the 0-th Fourier coefficient $\widehat{\zeta}(0) = 0$.

The problems (3.6) and (5.4) are equivalent for small solutions. Indeed, if $\varrho(t)$ is a solution of (3.6) corresponding to an initial value ϱ_0 which satisfies (5.3), then $\operatorname{vol}\Omega_{\varrho(t)} = \pi$ for all $t \in [0,T]$, meaning that

$$\int_{\mathbb{S}} (2\varrho(t) + \varrho^2(t))\, dx = 0 \quad \text{for all} \quad t \in [0,T].$$

Hence $\zeta(t)$ is a solution of (5.4). The converse is also true, that is if $\zeta(t)$ is a solution of (5.4), then $\varrho(t) := \sqrt{1 + 2\zeta(t)} - 1$ is a solution of (3.6) satisfying (5.3) for all $t \in [0,T]$. The next theorem describes the stability of the zero solution for (5.4) by using the equivalence of (3.6) and (5.4).

Theorem 5.1.2 *Let*

$$\iota_* := \frac{\omega}{\sqrt{2K}}.$$

(i) *If $\iota > \iota_*$, then the solution to (3.6) is exponentially stable. More precisely, given $C \in (0, (2K\iota^2 - \omega^2)/\overline{\mu}(0))$ there exist positive constants M and δ such that, if $\|\varrho_0\|_{C^{4+\alpha}(\mathbb{S})} \leq \delta$ and $\operatorname{vol}\Omega_{\varrho_0} = \pi$, then the solution to (3.6), corresponding to ϱ_0, exists in the large and*

$$\|\varrho(t)\|_{C^{4+\alpha}(\mathbb{S})} + \|\varrho'(t)\|_{C^{1+\alpha}(\mathbb{S})} \leq Me^{-Ct}\|\varrho_0\|_{C^{4+\alpha}(\mathbb{S})} \quad \text{for all } t \geq 0.$$

(ii) *If $\iota < \iota_*$, then the zero solution is unstable.*

(iii) *Let $\iota = \iota_*$ and $l \geq 2$ be fixed. Given $C \in (0, 6\gamma/\overline{\mu}(0))$, there exist positive constants M_l and δ_l such that, if $\|\varrho_0\|_{C^{4+\alpha}(\mathbb{S})} \leq \delta_l$, ϱ_0 is $2\pi/l$−periodic and additionally $\operatorname{vol}\Omega_{\varrho_0} = \pi$, then the solution to (3.6), corresponding to ϱ_0, exists in the large and*

$$\|\varrho(t)\|_{C^{4+\alpha}(\mathbb{S})} + \|\varrho'(t)\|_{C^{1+\alpha}(\mathbb{S})} \leq M_l e^{-Ct}\|\varrho_0\|_{C^{4+\alpha}(\mathbb{S})} \quad \text{for all } t \geq 0.$$

Proof Let $\iota > 0$ be given. Since $\Phi(\iota, 0) = 0$, the chain rule yields that

$$\partial_\zeta \Psi(\iota, 0)[\zeta] = \partial_\varrho \Phi(\iota, 0)[\zeta],$$

for all $\zeta \in h_0^{4+\alpha}(\mathbb{S})$, and so

$$\partial_\varrho \Psi(\iota, 0) \left[\sum_{k \in \mathbb{Z}\setminus\{0\}} \widehat{\zeta}(k) x^k \right] = \sum_{k \in \mathbb{Z}\setminus\{0\}} \lambda_k(\iota) \widehat{\zeta}(k) x^k \qquad (5.5)$$

for $\zeta = \sum_{k \in \mathbb{Z}\setminus\{0\}} \widehat{\zeta}(k) x^k \in h_0^{4+\alpha}(\mathbb{S})$, where $(\lambda_k(\iota))$ are given by (4.10). From (4.11) we deduce that $\lambda_k(\iota) \in \sigma(\partial_\varrho \Phi(\iota, 0))$ for all $k \in \mathbb{Z}$. Furthermore, the embedding $h^{4+\alpha}(\mathbb{S}) \hookrightarrow h^{1+\alpha}(\mathbb{S})$ is compact, hence $\partial_\varrho \Phi(\iota, 0)$ is an operator with compact resolvent. We infer from [45, Theorem III.8.29] that its spectrum consists entirely of isolated eigenvalues with finite multiplicity, thus

$$\sigma(\partial_\varrho \Phi(\iota, 0)) = \{\lambda_k(\iota) : k \in \mathbb{Z}\}.$$

For the restriction $\partial_\zeta \Psi(\iota, 0)$ we have then: $\sigma(\partial_\zeta \Psi(\iota, 0)) = \{\lambda_k(\iota) : k \in \mathbb{Z} \setminus \{0\}\}$.

Let us now notice that if $\iota < \iota_*$, then at least the eigenvalue $\lambda_1(\iota) = (\omega^2 - 2K\iota^2)/\overline{\mu}(0)$ is positive. The assumptions of [51, Theorem 9.1.3] are satisfied, meaning that the zero solution for problem (5.4) is unstable. The result holds then also for the zero solution of (3.6).

We prove now (i). One can easily check that, if $\iota > \iota_*$, then $\lambda_k(\iota) \leq -(2K\iota^2 - \omega^2)/\overline{\mu}(0)$ for all $k \in \mathbb{Z} \setminus \{0\}$, so that the spectrum of the linearisation $\partial_\zeta \Psi(\iota, 0)$ is bounded away from 0 in the left half of the complex plane. The principle of linearised stability applies then to this situation, meaning that the trivial solution of (5.4) is exponentially stable, i.e. given $C \in (0, (2K\iota^2 - \omega^2)/\overline{\mu}(0))$ there exist positive constants $\widetilde{M}$ and δ such that, if $\|\zeta_0\|_{C^{4+\alpha}(\mathbb{S})} \leq \delta$, then the solution to (5.4) satisfying initially $\zeta(0) = \zeta_0$, exists in the large and

$$\|\zeta(t)\|_{C^{4+\alpha}(\mathbb{S})} + \|\zeta'(t)\|_{C^{1+\alpha}(\mathbb{S})} \leq \widetilde{M} e^{-Ct} \|\zeta_0\|_{C^{4+\alpha}(\mathbb{S})} \quad \text{for all } t \geq 0.$$

Since $\varrho(t) = \sqrt{1 + 2\zeta(t)} - 1$, the desired assertion stated in (ii) is now obvious.

Finally, we consider (iii). Let $l \geq 2$ be fixed. Given $m \in \mathbb{N}$ and $\beta \in (0, 1)$, we define

$$h_l^{m+\beta}(\mathbb{S}) := \{\varrho \in h^{m+\beta}(\mathbb{S}) : \varrho(x) = \varrho(xe^{i2\pi/l}) \quad \text{for all} \quad x \in \mathbb{S}\},$$

the subspace of $h^{m+\beta}(\mathbb{S})$ consisting only of $2\pi/l$ periodic functions. Given $y \in \mathbb{R}^2$, $ye^{i2\pi/l}$ is the vector obtained by rotating y around the origin by the angle $2\pi/l$ in trigonometric sense, i.e.

$$ye^{i2\pi/l} = A(2\pi/l)y, \quad \text{where} \quad A(2\pi/l) := \begin{bmatrix} \cos(2\pi/l) & -\sin(2\pi/l) \\ \sin(2\pi/l) & \cos(2\pi/l) \end{bmatrix}.$$

We begin by showing that if $\varrho \in h_l^{4+\alpha}(\mathbb{S})$, then $\Phi(\iota_*, \varrho) \in h_l^{1+\alpha}(\mathbb{S})$. Indeed, if $\varrho \in h_l^{4+\alpha}(\mathbb{S})$ and $y \in \Omega_\varrho$, then $ye^{i2\pi/l} \in \Omega_\varrho$, since

$$|ye^{i2\pi/l}| = |y| < 1 + \varrho\left(\frac{y}{|y|}\right) = 1 + \varrho\left(\frac{ye^{i2\pi/l}}{|ye^{i2\pi/l}|}\right).$$

Letting $u = \Theta_*^\varrho \mathcal{T}(\iota_*, \varrho)$, denote the solution of the Dirichlet problem (4.1), the mapping

$$\bar{u} : \Omega_\varrho \to \mathbb{R}, \quad \bar{u}(y) = u(ye^{i2\pi/l}) \quad \text{for} \quad y \in \Omega_\varrho,$$

is well-defined and is also a solution of (4.1). The matrix $A(2\pi/l)$ is orthogonal and $\nabla u(y) = (A(2\pi/l))^T \nabla u(ye^{i2\pi/l})$, $y \in \Omega_\varrho$, where $(A(2\pi/l))^T$ is the transpose of $A(2\pi/l)$, so that

$$\mathcal{B}(\varrho, \mathcal{T}(\iota_*, \varrho))(xe^{i2\pi/l}) = -\frac{1}{\bar{\mu}(|\nabla u|^2)} \langle \nabla u | \nabla N_\varrho \rangle (\Theta_\varrho(xe^{i2\pi/l}))$$

$$= -\frac{1}{\bar{\mu}(|\nabla u|^2(\Theta_\varrho(x)e^{i2\pi/l}))} \langle \nabla u(\Theta_\varrho(x)e^{i2\pi/l}) | \nabla N_\varrho(\Theta_\varrho(xe^{i2\pi/l})) \rangle$$

$$= -\frac{1}{\bar{\mu}(|\nabla u|^2)} \langle A(2\pi/l)\nabla u | A(2\pi/l)\nabla N_\varrho \rangle (\Theta_\varrho(x))$$

$$= -\frac{1}{\bar{\mu}(|\nabla u|^2)} \left((A(2\pi/l)\nabla N_\varrho)^T A(2\pi/l)\nabla u\right)(\Theta_\varrho(x))$$

$$= -\frac{1}{\bar{\mu}(|\nabla u|^2)} \left((\nabla N_\varrho)^T (A(2\pi/l))^T A(2\pi/l)\nabla u\right)(\Theta_\varrho(x))$$

$$= -\frac{1}{\bar{\mu}(|\nabla u|^2)} \langle \nabla u | \nabla N_\varrho \rangle (\Theta_\varrho(x))$$

$$= \mathcal{B}(\varrho, \mathcal{T}(\iota_*, \varrho))(x).$$

We have thus shown that if $\varrho \in h_l^{4+\alpha}(\mathbb{S})$, then also $\Phi(\iota_*, \varrho) \in h_l^{1+\alpha}(\mathbb{S})$. The subspace

$$h_{0,l}^{m+\beta}(\mathbb{S}) := h_l^{m+\beta}(\mathbb{S}) \cap h_0^{m+\beta}(\mathbb{S}), \, m \in \mathbb{N} \quad \text{and } \beta \in (0,1),$$

consists of the functions $\zeta \in h^{m+\alpha}(\mathbb{S})$ with Fourier expansion

$$\zeta = \sum_{k \in \mathbb{Z}\setminus\{0\}} \widehat{\zeta}(kl) x^{kl}.$$

Taking into consideration that $\zeta = (2\varrho + \varrho^2)/2 \in h_l^{4+\alpha}(\mathbb{S})$ if $\varrho \in h_l^{4+\alpha}(\mathbb{S})$, we conclude that $\Psi \in C^\omega((0, \infty) \times \mathcal{V}_{0,l}, h_{0,l}^{1+\alpha}(\mathbb{S}))$, where $\mathcal{V}_{0,l} := \mathcal{V}_0 \cap h_l^{4+\alpha}(\mathbb{S})$. Hence, the spectrum of the Fréchet derivative $\partial_\zeta \Psi(\iota_*, 0)$ considered as operator in $\mathcal{L}(h_{0,l}^{4+\alpha}(\mathbb{S}), h_{0,l}^{1+\alpha}(\mathbb{S}))$ is $\sigma(\partial_\zeta \Psi(\iota_*, 0)) = \{\lambda_{kl}(\iota_*) : k \in \mathbb{Z} \setminus \{0\}\}$. Notice that $\sigma(\partial_\zeta \Psi(\iota_*, 0))$ consists only of negative eigenvalues and is bounded away from 0 in the left half of the complex plane since $\lambda_{kl}(\iota_*) \leq \lambda_2(\iota_*) = -6\gamma/\overline{\mu}(0)$. The assertion follows from the principle of linearised stability in the same manner as the assertion (i).

□

It is worth noticing the regularising effects of surface tension, viscosity and current's intensity. If the current's intensity and the surface tension of the interface separating the ferrofluid from air are large, and the viscosity $\mu(0)$ of the fluid at rest is small, then the fluid converges more rapidly to the unit disc.

5.2 Local bifurcation analysis

We noticed in Observation 5.1.1 that the stationary solutions of (3.6) are the periodic solutions of the ordinary differential equation (5.2). This is due to the fact that if ϱ is a stationary solution of (5.1), then by Stokes' theorem u, the potential in the domain Ω_ϱ, is constant also in the spatial variable. Recall that u is determined by the pair (ι, ϱ). This is the reason why we call from now on (ι, ϱ) stationary solution of (3.6).

In this section we use a bifurcation argument, with the current's intensity ι as bifurcation parameter and ω, γ fixed, to prove that there exist stationary fingering patterns that is nontrivial, finger shaped stationary solutions to our model. We shall refer to the set $\Sigma := \{(\iota, 0) : \iota \in (0, \infty)\}$ as the trivial branch of stationary solution of problem (5.1). Further on, we look for $\iota > 0$ with the property that $(\iota, 0) \in \Sigma$ is a bifurcation point from the trivial solution. The nontrivial stationary solutions are obtained by making use of the following theorem due to Crandall-Rabinowitz.

Theorem 5.2.1 (Bifurcation from simple eigenvalues) *Let X, Y be real Banach spaces and $G(\lambda, u)$ be a C^q ($q \geq 3$) mapping from a neighbourhood of a point $(\lambda_0, u_0) \in \mathbb{R} \times X$ into Y. Let the following assumptions hold:*

(i) $G(\lambda_0, u_0) = 0$, $\partial_\lambda G(\lambda_0, u_0) = 0$,

(ii) $\operatorname{Ker} \partial_u G(\lambda_0, u_0)$ *is one-dimensional, spanned by* v_0,

(iii) $\operatorname{Im} \partial_u G(\lambda_0, u_0)$ *has codimension 1,*

(iv) $\partial_\lambda \partial_\lambda G(\lambda_0, u_0) \in \operatorname{Im} \partial_u G(\lambda_0, u_0)$, $\partial_\lambda \partial_u G(\lambda_0, u_0) v_0 \notin \operatorname{Im} \partial_u G(\lambda_0, u_0)$.

Then (λ_0, u_0) is a bifurcation point of the equation

$$G(\lambda, u) = 0 \tag{5.6}$$

in the following sense: In a neighbourhood of (λ_0, u_0) the set of solutions of equation (5.6) consists of two C^{q-2} curves Σ_1 and Σ_2, which intersect only at the point (λ_0, u_0). Furthermore, Σ_1, Σ_2 can be parameterised as follows:

$\Sigma_1 : (\lambda, u(\lambda)), |\lambda - \lambda_0|$ *is small*, $u(\lambda_0) = u_0$, $u'(\lambda_0) = 0$,

$\Sigma_2 : (\lambda(\varepsilon), u(\varepsilon)), |\varepsilon|$ *is small*, $(\lambda(0), u(0)) = (\lambda_0, u_0)$, $u'(0) = v_0$.

Moreover, if Φ is analytic, then so are Σ_1 and Σ_2.

In view of relation (3.6), when determining the stationary solutions (ι, ϱ) of (3.6) corresponding to a fluid volume equal to π, that is the solutions of the free boundary problem (5.1) which additionally satisfy (5.3), we must not restrict to require that $\varrho \in \mathcal{V}$. Indeed, we have:

Proposition 5.2.2 *Let $(\iota, \varrho) \in (0, \infty) \times C^{2+\alpha}(\mathbb{S})$ satisfy $\varrho > -1$,*

$$\gamma \mathcal{K}(\varrho) - K\iota^2 \frac{1}{(1+\varrho)^2} - \frac{\omega^2}{2}(1+\varrho)^2 = const., \quad \text{and} \quad \int_\mathbb{S} 2\varrho + \varrho^2 \, dx = 0. \tag{5.7}$$

Then (ι, ϱ) is a stationary solution of (3.6).

Proof The first equation of (5.7) is obtain just by rewriting (5.2), hence, in view of Observation 5.1.1, we get that $\varrho \in C^\infty(\mathbb{S})$. Whence, by its definition, Ω_ϱ is a smooth domain. Denote by u the constant from the right hand side of the first equation of (5.7). It is obvious that (ϱ, u) is a solution of (5.1).

However, since (2.25) are fulfilled, u is uniquely determined by (ι, ϱ), and we are done. □

We transform now problem of finding the solutions of (5.7) into an operator equation such that the theorem on bifurcation from simple eigenvalues (Theorem 5.2.1) may be applied. It is convenient to incorporate the condition (5.3) in the spaces we work with. Setting $\zeta := 2\varrho + \varrho^2$, we are left to determine the 2π-periodic solution of the problem

$$\gamma \mathcal{K}(\sqrt{1+\zeta} - 1) - K\iota^2 \frac{1}{1+\zeta} - \frac{\omega^2}{2}(1+\zeta) = const. \quad \text{and} \quad \int_{\mathbb{S}} \zeta \, dx = 0, \quad (5.8)$$

which satisfy $\zeta > -1$. This is due to the fact that the mapping

$$\{\varrho \in C^{2+\alpha}(\mathbb{S}) : \varrho > -1\} \ni \varrho \mapsto \zeta = 2\varrho + \varrho^2 \in \{\zeta \in C^{2+\alpha}(\mathbb{S}) : \zeta > -1\}$$

is a bijection. We get rid now of the constant from the right hand side of the first equation of (5.8) by simply differentiating this equation with respect to the spatial variable x. This may be done since all the solutions of (5.8) are smooth. We define

$$\mathcal{W} := \{\zeta \in C^{3+\alpha}_{0,e}(\mathbb{S}) : \zeta > -1\}.$$

The space $C^{3+\alpha}_{0,e}(\mathbb{S})$ consists of the even functions of $C^{3+\alpha}(\mathbb{S})$ having integral mean zero,

$$C^{3+\alpha}_{0,e}(\mathbb{S}) := \left\{\zeta \in C^{3+\alpha}(\mathbb{S}) : \zeta \text{ is even and } \int_{\mathbb{S}} \zeta \, dx = 0\right\}.$$

Functions $\zeta \in C^{3+\alpha}_{0,e}(\mathbb{S})$ can be uniquely expanded, $\zeta = \sum_{k=1}^{\infty} a_k \cos(ks)$, where $a_k = 2\hat{\zeta}(k)$ for all $k \in \mathbb{N}, k \geq 1$. With these notation, finding the even stationary solutions of (ι, ϱ) of (3.6) satisfying (5.3) is equivalent with determining the solutions $(\iota, \zeta) \in (0, \infty) \times \mathcal{W}$ of the operator equation

$$\Upsilon(\iota, \zeta) = 0, \quad (5.9)$$

where $\Upsilon : (0, \infty) \times \mathcal{W} \subset \mathbb{R} \times C^{3+\alpha}_{0,e}(\mathbb{S}) \to C^{\alpha}_o(\mathbb{S})$ is the mapping defined by

$$\Upsilon(\iota, \zeta) := \left(\gamma \mathcal{K}(\sqrt{1+\zeta} - 1) - K\iota^2 \frac{1}{1+\zeta} - \frac{\omega^2}{2}(1+\zeta)\right)'. \quad (5.10)$$

The Banach space $C_o^\alpha(\mathbb{S})$ is the subspace of $C^\alpha(\mathbb{S})$ consisting only of odd functions. Taking into consideration that

$$\mathcal{K}(\sqrt{1+\zeta}-1) = \sqrt{1+\zeta}\frac{(1+\zeta)^2 + \frac{3}{4}\zeta'^2 - \frac{1}{2}(1+\zeta)\zeta''}{\left((1+\zeta)^2 + \frac{3}{4}\zeta'^2\right)^{3/2}}, \quad \zeta \in \mathcal{W},$$

we obtain that

$$\Upsilon(\iota,\zeta) = \gamma\frac{\zeta'}{2\sqrt{1+\zeta}}\frac{(1+\zeta)^2 + \frac{3}{4}\zeta'^2 - \frac{1}{2}(1+\zeta)\zeta''}{\left((1+\zeta)^2 + \frac{3}{4}\zeta'^2\right)^{3/2}} + K\iota^2\frac{\zeta'}{(1+\zeta)^2} - \frac{\omega^2}{2}\zeta'$$

$$+ \gamma\sqrt{1+\zeta}\frac{2(1+\zeta)\zeta' + \zeta'\zeta'' - \frac{1}{2}(1+\zeta)\zeta'''}{\left((1+\zeta)^2 + \frac{3}{4}\zeta'^2\right)^{3/2}}$$

$$- \frac{3}{2}\gamma\sqrt{1+\zeta}\left(2(1+\zeta)\zeta' + \frac{3}{2}\zeta'\zeta''\right)\frac{(1+\zeta)^2 + \frac{3}{4}\zeta'^2 - \frac{1}{2}(1+\zeta)\zeta''}{\left((1+\zeta)^2 + \frac{3}{4}\zeta'^2\right)^{5/2}}.$$

Since the derivative of an even function is odd we obtain that Υ is a well-defined function. Moreover, $\Upsilon \in C^\omega((0,\infty) \times \mathcal{W}, C_o^\alpha(\mathbb{S}))$. By the chain rule, the Fréchet derivative

$$\partial_\zeta \Upsilon(\iota,0)[\zeta] = \left(-\frac{\gamma}{2} + K\iota^2 - \frac{\omega^2}{2}\right)\zeta' - \frac{\gamma}{2}\zeta''',$$

hence

$$\partial_\zeta \Upsilon(\iota,0)\left[\sum_{k=1}^\infty a_k \cos(ks)\right] = \sum_{k=1}^\infty \left[\left(\frac{\gamma}{2} - K\iota^2 + \frac{\omega^2}{2}\right)k - \frac{\gamma}{2}k^3\right]a_k \sin(ks)$$

for all $\zeta = \sum_{k=1}^\infty a_k \cos(ks) \in C_{0,e}^{3+\alpha}(\mathbb{S})$. To simplify our notation, we set

$$\mu_k(\iota) := \left(\frac{\gamma}{2} - K\iota^2 + \frac{\omega^2}{2}\right)k - \frac{\gamma}{2}k^3$$

for all $\iota \in (0,\infty)$ and $k \in \mathbb{Z}$. We notice that, $\partial_\zeta \Upsilon(\iota,0)$ is the restriction to $C_{0,e}^{3+\alpha}(\mathbb{S})$ of the Fourier multiplication operator

$$A: C^{3+\alpha}(\mathbb{S},\mathbb{C}) \to C^\alpha(\mathbb{S},\mathbb{C}), \quad A\left[\sum_{k\in\mathbb{Z}} \widehat{\zeta}(k)x^k\right] = \sum_{k\in\mathbb{Z}} M_k(\iota)\widehat{\zeta}(k)x^k,$$

where

$$M_k(\iota) := \begin{cases} -\mu_k(\iota)i & , \; k \neq 0, \\ 1 & , \; k = 0, \end{cases}$$

and i is the root of -1. One can easily check that A is well-defined since the general Marcinkiwicz conditions (the assumptions $(i)-(iii)$ of Theorem 2.2.1) are satisfied by $(M_k(\iota))_{k\in\mathbb{Z}}$ when $s = 3+\alpha$ and $r = 1+\alpha$. Moreover, A maps real-valued functions onto real-valued functions. Given $\zeta \in C_{0,e}^{3+\alpha}(\mathbb{S})$, we obtain in view of $\widehat{\zeta}(-k) = \widehat{\zeta}(k) = a_k/2$, that

$$M_k(\iota)\widehat{\zeta}(k)e^{iks} + M_{-k}(\iota)\widehat{\zeta}(-k)e^{-iks} =$$
$$= M_k(\iota)\widehat{\zeta}(k)\cos(ks) + iM_k(\iota)\widehat{\zeta}(k)\sin(ks)$$
$$- M_k(\iota)\widehat{\zeta}(k)\cos(ks) + iM_k(\iota)\widehat{\zeta}(k)\sin(ks) = iM_k(\iota)a_k\sin(ks)$$
$$= \mu_k(\iota)a_k\sin(ks).$$

Hence, $\partial_\zeta \Upsilon(\iota, 0)$ is indeed the restriction of A to $C_{0,e}^{3+\alpha}(\mathbb{S})$.

We come to the main theorem of this part, which states that there exist besides the trivial circular equilibrium also nontrivial stationary solutions of the problem (3.6). We remark first that the set Σ defined at the beginning of this section can also be seen as the branch of trivial solutions of (5.9). For any integer $l \in \{1, \ldots, l_*\}$ we set

$$\bar{\iota}_l := \sqrt{\frac{\gamma + \omega^2 - \gamma l^2}{2K}}, \tag{5.11}$$

where we defined

$$l_* := \max\{l \in \mathbb{N} : \gamma + \omega^2 - \gamma k^2 > 0 \quad \text{for all} \quad 1 \leq k \leq l\}. \tag{5.12}$$

Obviously $l_* \geq 1$. Near $(\bar{\iota}_l, 0)$, $1 \leq l \leq l_*$, the set of equilibria of (5.9) consists of two analytic curves Σ and Σ_l which intersect in $(\bar{\iota}_l, 0)$. More exactly, we state:

Theorem 5.2.3 (Fingering patterns) *Given $1 \leq l \leq l_*$, the pair $(\bar{\iota}_l, 0)$ with $\bar{\iota}_l$ defined by relation (5.11), is a bifurcation point of the flat solution $(\iota, 0)$. More precisely, there exists $\delta > 0$ and a real analytical function $(\iota_l, \zeta_l) : (-\delta, \delta) \to (0, \infty) \times \mathcal{W}$, such that the pair $(\iota_l(\varepsilon), \zeta_l(\varepsilon))$ is a solution*

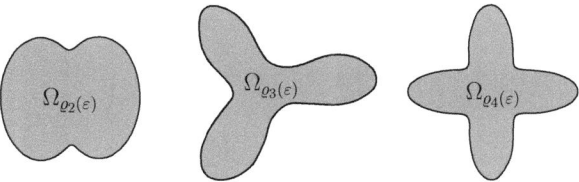

Figure 5.1: Possible stationary fingering patterns $\Omega_{\varrho_l(\varepsilon)}$ with $l \in \{2, 3, 4\}$.

of the equation (5.9) for all $\varepsilon \in (-\delta, \delta)$. The pair (ι_l, ζ_l) has the following asymptotic expressions for $\varepsilon \to 0$:

$$\iota_l(\varepsilon) = \bar{\iota}_l + O(\varepsilon^2),$$
$$\zeta_l(\varepsilon) = \varepsilon \cos(ls) + O(\varepsilon^2).$$

Moreover, any other ι is not a bifurcation point.

Proof Consider first the case when $\iota \notin \{\bar{\iota}_l : 1 \leq l \leq l_*\}$, meaning that $\mu_k(\iota) \neq 0$ for all $k \geq 1$. Form Theorem 2.2.1 we obtain then that A is an isomorphism. One can construct its inverse, which is also a Fourier multiplication operator as we did in the proof of Theorem 4.2.1. Thus, its restriction $\partial_\zeta \Upsilon(\iota, 0)$ has this property too, i.e. $\partial_\zeta \Upsilon(\iota, 0) \in \mathcal{L}is(C_{0,e}^{3+\alpha}(\mathbb{S}), C_o^\alpha(\mathbb{S}))$. We infer from the implicit function theorem that $(\iota, 0)$ is not a bifurcation point in this case and the stationary solutions of (3.6), close to $(\iota, 0)$, belong all to the trivial branch Σ.

If $\iota = \bar{\iota}_l$, for some $1 \leq l \leq l_*$, then $\mu_k(\iota) = 0$ iff $k = l$, hence

$$\operatorname{Ker} \partial_\zeta \Upsilon(\bar{\iota}_l, 0) = \operatorname{Span}\{\cos(ls)\} \quad \text{and} \quad \operatorname{codim} \operatorname{Im} \partial_\zeta \Upsilon(\bar{\iota}_l, 0) = 1.$$

Moreover, in view of

$$\partial_\iota \partial_\zeta \Upsilon(\bar{\iota}_l, 0) \left[\sum_{k \geq 1} a_k \cos(ks) \right] = -\sum_{k \geq 1} 2K \bar{\iota}_l k a_k \sin(ks),$$

we obtain that $\partial_\iota \partial_\zeta \Upsilon(\bar{\iota}_l, 0)[\cos(ls)] \notin \operatorname{Im} \partial_\zeta \Psi(\bar{\iota}_l, 0)$. Theorem (5.2.1) applies then to this situation and we obtain the desired result.

We are left to prove that $\iota'_l(0) = 0$. Therefore, it is convenient to write $\zeta_l(\varepsilon) = \varepsilon \cos(ls) + \sigma_l(\varepsilon)$, with $\sigma_l(0) = \sigma'_l(0) = 0$. Moreover, σ_l takes values in the closed complement of $\operatorname{Ker} \partial_\zeta \Upsilon(\bar\iota_l, 0)$ in $C^{3+\alpha}_{0,e}(\mathbb{S})$. Differentiating the relation $\Upsilon(\iota_l(\varepsilon), \varepsilon \cos(ls) + \sigma(\varepsilon)) = 0$ twice with respect to the variable ε, yields

$$2\iota'_l(0)\partial_\iota\partial_\zeta\Upsilon(\bar\iota_l,0)[\cos(ls)] + \partial_\zeta\partial_\zeta\Upsilon(\bar\iota_l,0)[\cos(ls)]^2 + \partial_\zeta\Upsilon(\bar\iota_l,0)[\sigma''_l(0)] = 0.$$

The last term of the sum belongs to the complement of $\mathbb{R} \cdot \{\sin(ls)\}$ in $C^\alpha_o(\mathbb{S})$. Hence, if we multiply the equality with $\sin(ls)$, and integrate over the unit circle, we obtain, after determining the derivative $\partial_\zeta\partial_\zeta\Upsilon(\bar\iota_l,0)[\cos(ls)]^2$, that

$$2lK\langle \sin(ls)|\sin(ls)\rangle \iota'_l(0) = \left[\frac{3}{4}\gamma l^3 - l\left(\frac{3}{4}\gamma - 2\bar\iota_l K\right)\right] \langle \sin(2ls)|\sin(ls)\rangle = 0,$$

and the proof is completed. $\square$

Possible stationary fingering pattern solutions $\varrho_l(\varepsilon) = \sqrt{1 + \zeta_l(\varepsilon)} - 1$ of problem (3.6) are pictured in Figure 5.1 in the case when $l_* \geq 4$. It is clear now why we defined Υ to act between the subspace $C^{3+\alpha}_{0,e}(\mathbb{S})$ and $C^\alpha_o(\mathbb{S})$. As mapping $\Upsilon \in C^\omega(C^{3+\alpha}_0(\mathbb{S}), C^\alpha_0(\mathbb{S}))$, the eigenspace of $\partial_\zeta\Upsilon(\bar\iota_l, 0)$ corresponding to the eigenvalue 0 is two-dimensional since it contains both functions x^l and x^{-l}, and an application of Theorem 5.2.1 would not have been possible.

5.3 Global bifurcation for analytic operators

Let $1 \leq l \leq l_*$ be given, where l_* is the positive integer defined by (5.12). We are interested to determine whether the local bifurcation branch (ι_l, ζ_l), obtained in Theorem 5.2.3, can be extended to the whole real line. Therefore, we first introduce some notation. We let Ξ denote the set of all solutions of (5.9) in $(0,\infty) \times \mathcal{W}$, and Σ^+_l be the restriction to $(0,\delta)$ of the positive bifurcation branch (ι_l, ζ_l) :

$$\Xi := \{(\iota, \zeta) \in (0,\infty) \times \mathcal{W} : \Upsilon(\iota, \zeta) = 0\},$$
$$\Sigma^+_l := \{(\iota_l(\varepsilon), \zeta_l(\varepsilon) : \varepsilon \in (0,\delta))\}.$$

In order to obtain the main result of this section, stated in Theorem 5.3.2, we need some preliminary results. We shall say that a set is regularly bounded in $(0,\infty) \times \mathcal{W}$ if it is both bounded in $\mathbb{R} \times C^{3+\alpha}_{0,e}(\mathbb{S})$ and bounded away from the boundary of $(0,\infty) \times \mathcal{W}$.

Lemma 5.3.1 *We have that $\Xi \subset (0,\infty) \times C^\infty(\mathbb{S})$. Moreover, the regularly bounded and closed subsets of Ξ are compact.*

Proof The smoothness of the equilibria has been already discussed in Observation 5.1.1. We are left to prove that the regularly bounded and closed subsets of Ξ are compact. This is equivalent to showing that any bounded sequence $((\iota_n, \zeta_n))_n$ in Ξ has a convergent subsequence. In view of $\Upsilon(\iota_n, \zeta_n) = 0$, we find for all $n \in \mathbb{N}$

$$\zeta_n''' = \frac{2\left((1+\zeta_n)^2 + \tfrac{3}{4}\zeta_n'^2\right)^{3/2}}{\gamma(1+\zeta_n)^{3/2}} \left[\gamma \frac{\zeta_n'}{2\sqrt{1+\zeta_n}} \frac{(1+\zeta_n)^2 + \tfrac{3}{4}\zeta_n'^2 - \tfrac{1}{2}(1+\zeta_n)\zeta_n''}{\left((1+\zeta_n)^2 + \tfrac{3}{4}\zeta_n'^2\right)^{3/2}} \right.$$
$$- \frac{3}{2}\gamma\sqrt{1+\zeta_n}\left(2(1+\zeta_n)\zeta_n' + \frac{3}{2}\zeta_n'\zeta_n''\right) \frac{(1+\zeta_n)^2 + \tfrac{3}{4}\zeta_n'^2 - \tfrac{1}{2}(1+\zeta_n)\zeta_n''}{\left((1+\zeta_n)^2 + \tfrac{3}{4}\zeta_n'^2\right)^{5/2}}$$
$$\left. + K\iota_n^2 \frac{\zeta_n'}{(1+\zeta_n)^2} - \frac{\omega^2}{2}\zeta_n' \right] + 4\zeta_n' + \frac{2\zeta_n'\zeta_n''}{1+\zeta_n}.$$

Since $((\iota_n, \zeta_n))_n$ is bounded away from the boundary of $(0,\infty) \times \mathcal{W}$, we get that $1 + \zeta_n \geq c > 0$ for all $n \in \mathbb{N}$, whence (ζ_n) is a bounded sequence in $C^{4+\alpha}(\mathbb{S})$. The embedding $C^{4+\alpha}(\mathbb{S}) \hookrightarrow C^{3+\alpha}(\mathbb{S})$ is compact, and (ζ_n) possesses therefore a convergent subsequence in $C^{3+\alpha}_{0,e}(\mathbb{S})$. Taking into consideration that (ι_n) is bounded too, we finished the proof. □

The main result of this section is the following global bifurcation theorem which is based on general results presented in [12]. It strongly relies on Lemma 5.3.1 and of the fact that

$$\partial_\zeta \Upsilon(\iota, \zeta) \text{ is a Fredholm operator of index } 0 \text{ for all } (\iota, \zeta) \in \Xi. \qquad (5.13)$$

Our analysis then shows that the assumptions of [12, Theorem 9.1.1] are all satisfied and we conclude that the local bifurcation branches obtained in Theorem 5.2.3 can de expended to the whole real line $\mathbb{R}$.

Theorem 5.3.2 *There exists a continuous curve Σ_l which extends Σ_l^+ as follows:*

(a) $\Sigma_l = \{(u_l(\varepsilon), \zeta_l(\varepsilon)) : \varepsilon \in [0, \infty)\} \subset (0, \infty) \times \mathcal{W}$ *where* $(u_l, \zeta_l) : [0, \infty) \to (0, \infty) \times \mathcal{W}$ *is continuous.*

(b) $\Sigma_l^+ \subset \Sigma_l \subset \Xi$;

(c) *The set* $\{\varepsilon \geq 0 : \operatorname{Ker} \partial_\zeta \Upsilon(u_l(\varepsilon), \zeta_l(\varepsilon)) \neq \{0\}\}$ *has no accumulation points.*

(d) *At each point, Σ_l has a local analytical re-parametrisation in the following sense. In a neighbourhood of $\varepsilon = 0$, Σ_l^+ and Σ_l coincide. For each $\varepsilon^* \in (0, \infty)$ there exists $\rho^* : (-1, 1) \to \mathbb{R}$ which is continuous, injective, and*

$$\rho^*(0) = \varepsilon^*, \quad t \mapsto (u_l(\rho^*(t)), \zeta_l(\rho^*(t))), \, t \in (-1, 1), \quad \text{is analytic.}$$

Furthermore u_l is injective on a right neighbourhood of 0, and for $\varepsilon^* > 0$ there exists $\delta^* > 0$ such that u_l is injective on $[\varepsilon^*, \varepsilon^* + \delta^*]$ and on $[\varepsilon^* - \delta^*, \varepsilon^*]$.

(e) *One of the following situations occurs:*

(i) $\|(u_l(\varepsilon), \zeta_l(\varepsilon))\|_{\mathbb{R} \times C^{3+\alpha}(S)} \to \infty$ *as* $\varepsilon \to \infty$.

(ii) $(u_l(\varepsilon), \zeta_l(\varepsilon))$ *approaches the boundary of* $(0, \infty) \times \mathcal{W}$ *as* $\varepsilon \to \infty$.

(iii) Σ_l *is a closed loop, i.e. for some* $T > 0$,

$$\Sigma_l = \{(u_l(\varepsilon), \zeta_l(\varepsilon)) : 0 \leq \varepsilon \leq T\},$$

and $(u_l(T), \zeta_l(T)) = (\bar{u}_l, 0)$. *We may assume that $T > 0$ is the smallest number with this property.*

(f) *If, for some $\varepsilon_1 \neq \varepsilon_2$, $(u_l(\varepsilon_1), \zeta_l(\varepsilon_1)) = (u_l(\varepsilon_2), \zeta_l(\varepsilon_2))$ where we have that $\operatorname{Ker} \partial_f \Phi(u_l(\varepsilon_1), \zeta_l(\varepsilon_1)) = \{0\}$, then $(e)(iii)$ occurs, and $|\varepsilon_1 - \varepsilon_2|$ is an integer multiple of T.*

To apply the result [12, Theorem 9.1.1] to our special situation we still have to check that the relation (5.13) is valid. Therefore, we prove the more general result:

Lemma 5.3.3 *Given $(\iota_0, \zeta_0) \in (0, \infty) \times \mathcal{W}$, the Fréchet derivative $\partial_\zeta \Upsilon(\iota_0, \zeta_0)$ is a Fredholm operator of index 0.*

Proof Let us first notice that the highest order term of the partial derivative $\partial_\zeta \Upsilon(\iota_0, \zeta_0)[\zeta]$, $\zeta \in C_{0,e}^{3+\alpha}(\mathbb{S})$, is the term containing the third derivative ζ'''. Consequently, we can write the derivative $\partial_\zeta \Upsilon(\iota_0, \zeta_0)$ as the sum

$$\partial_\zeta \Upsilon(\iota_0, \zeta_0) = A + B$$

where $A : C_{0,e}^{3+\alpha}(\mathbb{S}) \to C_o^{\alpha}(\mathbb{S})$ is the linear operator defined by

$$A\zeta := \sigma_0 \zeta''', \quad \zeta \in C_{0,e}^{3+\alpha}(\mathbb{S}),$$

and

$$\sigma_0 := -\frac{\gamma(1+\zeta_0)^{3/2}}{2\left((1+\zeta_0)^2 + \frac{3}{4}\zeta_0'^2\right)^{3/2}}.$$

The operator $B : C_{0,e}^{3+\alpha}(\mathbb{S}) \to C_o^{\alpha}(\mathbb{S})$ is given by the relation

$$B\zeta = \sigma_1 \zeta'' + \sigma_2 \zeta' + \sigma_3 \zeta,$$

where σ_j, $1 \leq j \leq 3$ belong to $C^\alpha(\mathbb{S})$. Of course, σ_1 and σ_3 are odd, and σ_2 is even. The operator B contains all the lower order terms of the derivative $\partial_\zeta \Upsilon(\iota_0, \zeta_0)[\zeta]$, and is therefore compact. Hence, $\partial_\zeta \Upsilon(\iota_0, \zeta_0)$ is the perturbation of A with the compact operator B. If we show that A is a Fredholm operator of Fredholm index $\operatorname{ind} A = 0$, then, in view of [39, Satz 81.3], the conclusion is immediate.

We show in the last part of the proof that A is not only a Fredholm operator of index 0, but even more, an isomorphism. Due to $\sigma_0 < 0$, it is clear that A is one-to-one, so that we are left to prove that A is also onto. Let $\varrho \in C_o^\alpha(\mathbb{S})$ be given. Our aim is to find $\zeta \in C_{0,e}^{3+\alpha}(\mathbb{S})$ such that $\sigma_0 \zeta''' = \varrho$. The function $\varrho_1 : \mathbb{R} \to \mathbb{R}$, with

$$\varrho_1(t) = \int_0^t \varrho/\sigma_0(s)\, ds - b \quad \text{and} \quad b := \frac{1}{2\pi} \int_0^{2\pi} \int_0^t \varrho/\sigma_0(s)\, ds\, dt,$$

is even, 2π–periodic and $\varrho_1' = \varrho/\sigma_0$. Indeed, taking into consideration that the ratio ϱ/σ_0 is odd, we obtain

$$\varrho_1(t+2\pi) = \int_0^{t+2\pi} \varrho/\sigma_0(s)\,ds - b = \varrho_1(t) + \int_t^{t+2\pi} \varrho/\sigma_0(s)\,ds$$

$$= \varrho_1(t) + \int_{-\pi}^{\pi} \varrho/\sigma_0(s)\,ds = \varrho_1(t),$$

and also

$$\varrho_1(-t) = \int_0^{-t} \varrho/\sigma_0(s)\,ds - b = -\int_0^t \varrho/\sigma_1(-s)\,ds - b = \varrho_1(t)$$

for all $t \in \mathbb{R}$. Furthermore, ϱ_1 has integral mean equal to 0, hence $\varrho_2 : \mathbb{R} \to \mathbb{R}$, with

$$\varrho_2(t) = \int_0^t \varrho_1(s)\,ds, \quad t \in \mathbb{R},$$

is an odd primitive of ϱ_1 belonging to $C^{2+\alpha}(\mathbb{S})$. Finally, setting

$$\zeta(t) := \int_0^t \varrho_2(s)\,ds - c, \quad \text{where} \quad c := \frac{1}{2\pi}\int_0^{2\pi}\int_0^t \varrho_2(s)\,ds\,dt,$$

we have found the desired function $\zeta \in C_{0,e}^{3+\alpha}(\mathbb{S})$, which solves the equation $\sigma_0 \zeta''' = \varrho$. □

Lastly, we give a lemma characterising the behaviour of the fingering patterns as the currents intensity tends to zero along a bifurcation branch. Namely, if the bifurcation solutions remain bounded, but u_l becomes very large, then the fingering patterns disappear in the sense that they converge to the rotationally invariant equilibrium. More exactly, we state:

Corollary 5.3.4 *Let Σ_l be the global bifurcation branch found in Theorem 5.3.2. If*

(a) $\lim_{\varepsilon \to \infty} u_l(\varepsilon) = \infty$,

(b) $\inf_{\varepsilon \geq 0} \zeta_l(\varepsilon) > -1$,

(c) $\sup_{\varepsilon \geq 0} \|\zeta_l(\varepsilon)\|_{C^{3+\alpha}(\mathbb{S})} < \infty$,

then $\lim_{\varepsilon \to \infty} \zeta_l(\varepsilon) = 0$ in $C^{3+\beta}(\mathbb{S})$ for all $\beta < \alpha$.

Proof Let $\beta < \alpha$ be given. We assume by contradiction that we found a strictly increasing sequence $(\varepsilon_n)_n$, converging to infinity, with the property that $\|\zeta_l(\varepsilon_n)\|_{C^{3+\beta}(\mathbb{S})} \geq \delta > 0$ for all $n \in \mathbb{N}$. We may choose $(\varepsilon_n)_n$ such that $\iota_l(\varepsilon_n) \geq 1$ for all $n \in \mathbb{N}$. Moreover, since $(\zeta_l(\varepsilon_n))_n$ is bounded in $C^{3+\alpha}(\mathbb{S})$ we presuppose that it converges $\zeta_l(\varepsilon_n) \to_{n\to\infty} \zeta$ in $C^{3+\beta}(\mathbb{S})$ and $\zeta \neq 0$.

By dividing with $\iota_l(\varepsilon_n)$ the relation $\Upsilon(\iota_l(\varepsilon_n), \zeta_l(\varepsilon_n)) = 0$, and letting $n \to \infty$ we obtain that $\zeta' = 0$. Since ζ has integral mean equal to zero, this yields $\zeta = 0$. We conclude that our assumption was false and the proof is completed.

$\square$

5.4 Global bifurcation via Leray-Schauder degree

In this last section we consider the situation when l_*, the constant defined by the relation (5.12), is $l_* = 1$. This is possible only if the angular velocity ω and the surface tension coefficient verify the inequation

$$\omega^2 \leq 3\gamma. \tag{5.14}$$

As a consequence of (5.14) we obtain that there exists a unique bifurcation point of the set Σ consisting of the trivial solutions of (5.9). Let $\mathfrak{S}$ be the closure of the set of nontrivial solutions of (5.9). We are interested what happens with the connected component $\mathcal{C}$ of $\mathfrak{S}$, which contains the continuous branch Σ_1 obtained in Theorem 5.3.2. More precisely, we prove that $\mathcal{C}$ is not regularly bounded in $(0, \infty) \times \mathcal{W}$. Particularly, $\mathcal{C}$ can not be a closed loop.

Theorem 5.4.1 *Assume that (5.14) is satisfied. The connected component $\mathcal{C}$ of $\mathfrak{S}$ to which $(\bar{\iota}_1, 0)$ belongs is not regularly bounded in $(0, \infty) \times \mathcal{W}$.*

To prove Theorem 5.4.1 we shall use the global Rabinowitz bifurcation theorem, which relies strongly on the Leray-Schauder degree. We shall apply its variant [46, Theorem II.3.3] (see Theorem 5.4.2). Before of that we introduce some notation.

Let $\mathbb{E}$ be a Banach space and $F : \mathbb{R} \times \mathbb{E} \to \mathbb{E}$ a smooth mapping. It is crucial for bifurcation at $(\iota_0, 0) \in \mathbb{R} \times \mathbb{E}$ how the eigenvalue 0 perturbs for $\partial_\zeta F(\iota_0, 0)$ when ι varies in a neighbourhood of ι_0. The 0−group of $\partial_\zeta F(\iota, 0)$ is the set consisting of the eigenvalues of $\partial_\zeta F(\iota, 0)$ near 0, cf. [46]. Moreover, the eigenvalues in the zero group depend continuously on ι. Define $\sigma^<(\iota) = 1$, if there are no negative real eigenvalues in the 0−group of $\partial_\zeta F(\iota, 0)$, and $\sigma^<(\iota) = (-1)^{m_1+\ldots+m_k}$ if $\mu_1, \ldots, \mu_k$ are all negative real eigenvalues in the 0-group having algebraic multiplicities $m_1, \ldots, m_k$, respectively. If

$$\partial_\zeta F(\iota, 0) \text{ is bijective for } \iota \in (\iota_0 - \delta, \iota_0) \cup (\iota_0, \iota_0 + \delta),$$
$$\text{and } \sigma^<(\iota) \text{ changes at } \iota = \iota_0, \tag{5.15}$$

then we say that $\partial_\zeta F(\iota, 0)$ has an odd crossing number at $\iota = \iota_0$. We use the following global bifurcation theorem:

Theorem 5.4.2 *Assume that $F \in C(\mathbb{R} \times \mathbb{E}, \mathbb{E})$, $F(\iota, \zeta) = \zeta + f(\iota, \zeta)$, where $f : \mathbb{R} \times \mathbb{E} \to \mathbb{E}$ is completely continuous, and $\partial_\zeta F(\iota, 0) = id_{\mathbb{E}} + \partial_\zeta f(\iota_0, \cdot) \in C(\mathbb{R}, \mathcal{L}(\mathbb{E}, \mathbb{E}))$. Let $\mathfrak{R}$ denote the closure of the set of nontrivial solutions of $F(\iota, \zeta) = 0$ in $\mathbb{R} \times \mathbb{E}$. Assume that $\partial_\zeta F(\iota, 0)$ has an odd crossing number at $\iota = \iota_0$. Then $(\iota_0, 0) \in \mathfrak{R}$, and let C be the (connected) component of $\mathfrak{R}$ to which $(\iota_0, 0)$ belongs. Then*

(i) *C is unbounded, or*

(ii) *C contains some $(\iota_1, 0)$, where $\iota_1 \neq \iota_0$.*

To make possible a direct application of this theorem we have to rewrite first the operator equation (5.9) in a more accessible way. We recall that the 0−level set of Υ, $\Upsilon^{-1}(0)$, is exactly the set consisting of all stationary solutions of (3.6). Using the relation found in the proof of Lemma 5.3.1, we write the equation (5.9) in the form

$$\zeta''' + h(\iota, \zeta) = 0, \tag{5.16}$$

where $h : (0,\infty) \times \mathcal{W} \subset (0,\infty) \times C^{3+\alpha}_{0,e}(\mathbb{S}) \to C^{\alpha}_o(\mathbb{S})$ is the mapping defined by

$$h(\iota,\zeta) = -\frac{2\left((1+\zeta)^2 + \frac{3}{4}\zeta'^2\right)^{3/2}}{\gamma(1+\zeta)^{3/2}}\left[\gamma\frac{\zeta'}{2\sqrt{1+\zeta}}\frac{(1+\zeta)^2 + \frac{3}{4}\zeta'^2 - \frac{1}{2}(1+\zeta)\zeta''}{\left((1+\zeta)^2 + \frac{3}{4}\zeta'^2\right)^{3/2}}\right.$$

$$-\frac{3}{2}\gamma\sqrt{1+\zeta}\left(2(1+\zeta)\zeta' + \frac{3}{2}\zeta'\zeta''\right)\frac{(1+\zeta)^2 + \frac{3}{4}\zeta'^2 - \frac{1}{2}(1+\zeta)\zeta''}{\left((1+\zeta)^2 + \frac{3}{4}\zeta'^2\right)^{5/2}}$$

$$\left. + K\iota^2\frac{\zeta'}{(1+\zeta)^2} - \frac{\omega^2}{2}\zeta'\right] - 4\zeta' - \frac{2\zeta'\zeta''}{1+\zeta}.$$

Furthermore, the proof of Lemma 5.3.3 shows that $A : C^{3+\alpha}_{0,e}(\mathbb{S}) \to C^{\alpha}_o(\mathbb{S})$, with $A\zeta := \zeta'''$ for $\zeta \in C^{3+\alpha}_{0,e}(\mathbb{S})$, is an isomorphism. By applying the inverse A^{-1} to the equation (5.16), we find out that determining the equilibria of problem (3.6) is equivalent to finding the solutions of the equation

$$G(\iota,\zeta) := \zeta + g(\iota,\zeta) = 0, \tag{5.17}$$

where $g : (0,\infty) \times \mathcal{W} \subset (0,\infty) \times C^{3+\alpha}_{0,e}(\mathbb{S}) \to C^{3+\alpha}_{0,e}(\mathbb{S})$ is the mapping defined by $g(\iota,\zeta) = A^{-1}h(\iota,\zeta)$ for $(\iota,\zeta) \in (0,\infty) \times \mathcal{W}$. With these notations we come to the proof of our main result.

Proof (Proof of Theorem 5.4.1) Assume by contradiction that the maximal connected component $\mathcal{C}$ of $\mathfrak{S}$ which contains $(\bar\iota_1, 0)$ is regularly bounded in $(0,\infty) \times \mathcal{W}$. Therefore it must be closed, hence by Lemma 5.3.1 is compact. There exist then an open and regularly bounded neighbourhood U of $\mathcal{C}$ in $(0,\infty) \times \mathcal{W}$ (see the Figure 5.2 below).

We extend now the mapping G on the whole $\mathbb{R} \times C^{3+\alpha}_{0,e}(\mathbb{S})$. Since U is bounded away from the boundary of $(0,\infty) \times \mathcal{W}$, we find a negative constant $\vartheta > -1$ such that $U \subset \{(\iota,\zeta) : \zeta > \vartheta\}$. Let $\eta \in C^{\infty}(\mathbb{R},[0,1])$ be a function satisfying

$$\eta(\zeta) = \begin{cases} 1 &, \zeta > \vartheta, \\ 0 &, \zeta < \overline{\vartheta}, \end{cases}$$

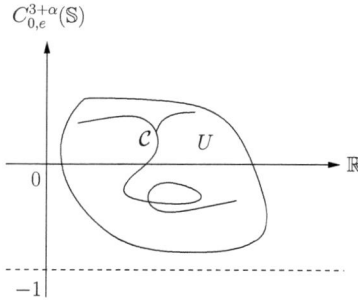

Figure 5.2: The neighbourhood U.

with $\vartheta > \bar{\vartheta} > -1$. Given $(\iota, \zeta) \in \mathbb{R} \times C_{0,e}^{3+\alpha}(\mathbb{S})$ we set $F(\iota, \zeta) = \zeta + f(\iota, \zeta)$, where

$$f(\iota,\zeta) = A^{-1} \left\{ -\frac{2\left((1+\eta(\zeta)\zeta)^2 + \frac{3}{4}\zeta'^2\right)^{3/2}}{\gamma(1+\eta(\zeta)\zeta)^{3/2}} \right.$$

$$\times \left[\gamma \frac{\zeta'}{2\sqrt{1+\eta(\zeta)\zeta}} \frac{(1+\zeta)^2 + \frac{3}{4}\zeta'^2 - \frac{1}{2}(1+\zeta)\zeta''}{\left((1+\eta(\zeta)\zeta)^2 + \frac{3}{4}\zeta'^2\right)^{3/2}} \right.$$

$$\left. - \frac{3\gamma}{2}\sqrt{1+\eta(\zeta)\zeta}\left(2(1+\zeta)\zeta' + \frac{3}{2}\zeta'\zeta''\right)\frac{(1+\zeta)^2 + \frac{3}{4}\zeta'^2 - \frac{1}{2}(1+\zeta)\zeta''}{\left((1+\eta(\zeta)\zeta)^2 + \frac{3}{4}\zeta'^2\right)^{5/2}} \right.$$

$$\left. + K\iota^2 \frac{\zeta'}{(1+\eta(\zeta)\zeta)^2} - \frac{\omega^2}{2}\zeta' \right] - 4\zeta' - \frac{2\zeta'\zeta''}{1+\eta(\zeta)\zeta} \right\}$$

Since $\eta(\zeta)\zeta \geq \bar{\vartheta} > -1$ for all $\zeta \in \mathbb{R}$, we have defined in this way a smooth mapping $F \in C^\infty(\mathbb{R} \times C_{0,e}^{3+\alpha}(\mathbb{S}), C_{0,e}^{3+\alpha}(\mathbb{S}))$, and moreover, $F = G$ in U. Let $\mathfrak{R}$ denote the closure in $\mathbb{R} \times C_{0,e}^{3+\alpha}(\mathbb{S})$ of the set consisting of the nontrivial solutions of the equation $F(\iota, \zeta) = 0$. The connected component of $\mathfrak{R}$ to which $(\bar{\iota}_1, 0)$ belongs must be, in view of $F = G$ in U, exactly $\mathcal{C}$.

We prove now that f is completely continuous, i.e. it maps bounded sets onto relatively compact sets. Indeed, if (ι_n, ζ_n) is a bounded sequence in $\mathbb{R} \times C_{0,e}^{3+\alpha}(\mathbb{S})$, then $Af(\iota_n, \zeta_n)$ is a bounded sequence in $C^{1+\alpha}(\mathbb{S})$. Hence, we find a convergent subsequence of $Af(\iota_n, \zeta_n)$ in $C_0^{\alpha}(\mathbb{S})$. Taking into consideration that A is an isomorphism we conclude that $f(\iota_n, \zeta_n)$ has a convergent subsequence in $C_{0,e}^{3+\alpha}(\mathbb{S})$. Since (ι_n, ζ_n) was arbitrarily chosen, this means that f is completely continuous.

In order to apply Theorem [46, Theorem II.3.3] we still have to check that $\partial_\zeta F(\iota, 0)$ has an odd crossing number at $\iota = \bar{\iota}_1$. In view of $\eta = 1$ on the set $\{\zeta : \zeta > \vartheta\}$, we have that $\partial_\zeta F(\iota, 0) = \partial_\zeta G(\iota, 0)$ for all $\iota > 0$. It is an elementary, though lengthy, computation which shows that

$$\partial_\zeta G(\iota, 0)[\zeta] = \zeta + \left(1 + \frac{\omega^2 - 2K\iota^2}{\gamma}\right) A^{-1} \zeta'$$

for all $\zeta \in C_{0,e}^{3+\alpha}(\mathbb{S})$. Since

$$A\left[\sum_{k=1}^{\infty} a_k \cos(ks)\right] = \sum_{k=1}^{\infty} k^3 a_k \sin(ks) \quad \text{for} \quad \zeta = \sum_{k=1}^{\infty} a_k \cos(ks) \in C_{0,e}^{3+\alpha}(\mathbb{S}),$$

we obtain for $\iota > 0$ that

$$\partial_\zeta F(\iota, 0)\left[\sum_{k=1}^{\infty} a_k \cos(ks)\right] = \sum_{k=1}^{\infty} m_k(\iota) a_k \cos(ks),$$

where

$$m_k(\iota) := 1 - \left(1 + \frac{\omega^2 - 2K\iota^2}{\gamma}\right) \frac{1}{k^2}$$

for all $k \in \mathbb{N}$, $k \geq 1$. By applying the Theorem 2.2.1 we see that the spectrum of $\partial_\zeta F(\iota, 0)$ consists only of eigenvalues

$$\sigma(\partial_\zeta F(\iota, 0)) = \sigma_p(\partial_\zeta F(\iota, 0)) = \{m_k(\iota) : k \geq 1\}.$$

Notice that $m_k(\iota) = 0$ iff $\iota = \bar{\iota}_k$, where $\bar{\iota}_k$ is the constant defined by (5.11). The condition (5.14) imposed at the beginning of the section ensures that $\bar{\iota}_k < 0$ for $k \geq 2$, so that if $m_k(\iota) = 0$ and $\iota > 0$, then it must hold $\iota = \bar{\iota}_1$ and $k = 1$.

Consequently, for $\iota \in (0, \bar{\iota}_1) \cup (\bar{\iota}_1, \infty)$ the derivative $\partial_\zeta F(\iota, 0)$ is an isomorphism. Moreover, for $\iota < \bar{\iota}_1$

$$m_1(\iota) = \frac{2K\iota^2 - \omega^2}{\gamma} < 0,$$

and

$$m_k(\iota) = \frac{(k^2 - 1)\gamma - \omega^2 + 2K\iota^2}{\gamma k^2} \geq \frac{3\gamma - \omega^2 + 2K\iota^2}{\gamma k^2} > 0,$$

hence $\sigma^<(\iota) = -1$ for all $\iota \in (0, \bar{\iota}_1)$. If $\iota > \bar{\iota}_1$, then $m_k(\iota) > 0$ for all $k \geq 1$, so that $\sigma^<(\iota) = 1$ for $\iota > \bar{\iota}_1$. Consequently, $\partial_\zeta F(\iota, 0)$ has an odd crossing number at $\bar{\iota}_1$, and the assumptions of Theorem 5.4.2 are all verified.

We conclude that $\mathcal{C}$ is unbounded or contains some $(\iota_1, 0)$ with $\bar{\iota}_1 \neq \iota_1$. However, by assumption, $\mathcal{C}$ is regularly bounded so that there must exist a bifurcation point $(\iota_1, 0) \in \Sigma$ for equation (5.9), which belongs to U. But condition (5.14) and Theorem 5.3.2 ensure that $(\bar{\iota}_1, 0)$ is the single bifurcation point of Σ. This completes the proof. □

Part II

Well-posedness, analytic dependence, and bifurcation analysis for the Muskat problem

Chapter 6

The mathematical model

We start this chapter by briefly describing the physical setting we are interested in. The mathematical model, a two-phase moving boundary problem, is transformed on a fixed reference domain by strengthening the moving interface between the fluids. Section 6.2 contains also the main existence, uniqueness, and analytic dependence on the initial data result, Theorem 6.2.2.

6.1 The mathematical model

The Muskat problem, proposed in 1934 (see [55]), describes the encroachment of water into an oil sand. We consider in our work a system consisting of two incompressible Newtonian fluids, which occupy together the entire void space of a porous medium or a vertical Hele-Shaw cell. These fluid phases are assumed to be separated from each other by a sharp interface which moves along with the fluids and we presuppose that there are equal quantities of both fluids in the cell. We shall refer to the fluid on the bottom of the cell (Fluid 1) using the subscript − and to the other one (Fluid 2) by using the subscript +. In our setting, the bottom of the cell and the laterals are taken to be impermeable.

The Hele-Shaw cell

$$\widetilde{\Omega} := (-2\pi, 2\pi) \times (-1, 1) \times (-b, b)$$

consists, at each time $t \geq 0$, of two wetting phases $\widetilde{\Omega}_-(t)$ and $\widetilde{\Omega}_+(t)$, corresponding to each of the fluids, and we presuppose that the distance

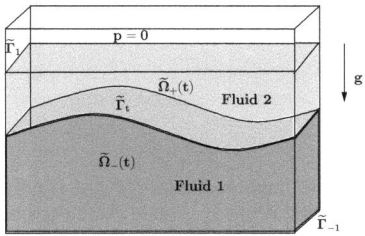

Figure 6.1: The Hele-Shaw cell.

between the plates is small compared to the length and height of the cell $b \ll 1$. The two phases are separated by the moving interface $\widetilde{\Gamma}(t)$. To ease our notations, we let the bottom $\widetilde{\Gamma}_{-1}$ to be the plane $[y = -1]$ and the boundary component $\widetilde{\Gamma}_1$ to be the plane $[y = 1]$. The situation is illustrated in Figure 6.1. We assume that the only external force acting on the fluid system is the gravity. Hence, we are in the situation considered in Subsection 2.1.6. By integrating over the small gap between the plates we obtain a two-dimensional model describing the physical setting. Namely, defining the potentials

$$u_\pm := p_\pm + g\rho_\pm y, \qquad (6.1)$$

in $\widetilde{\Omega}_\pm(t)$, we obtain, after some approximation, that $u_\pm$ do not depend on the gap coordinate z. We maintain in this part the notation from the first chapter of this work, namely we shall write $\rho_\pm$ to denote the densities of the fluids, $p_\pm$ for their pressure, and g is again the gravity constant. Moreover, the gap averaged velocities $\vec{v}_\pm$, defined by relation (2.28), satisfy Darcy's law

$$\vec{v}_\pm = -\frac{1}{\mu_\pm}\nabla u_\pm \quad \text{in} \quad \Omega_\pm(t), \qquad (6.2a)$$

where $\Omega_\pm(t)$ is the projection of $\widetilde{\Omega}_\pm(t)$ on $\mathbb{R}^2$, the plane parallel to the plates which contains the zero vector in $\mathbb{R}^3$. Similarly, we denote by Γ_1, Γ_{-1} and $\Gamma(t)$ the projection of $\widetilde{\Gamma}_1, \widetilde{\Gamma}_{-1}$, and $\widetilde{\Gamma}(t)$ on $\mathbb{R}^2$, respectively. Remind that $\mu_\pm$ are the viscosity constants of the two fluids. The relations (6.2a) have been obtain by normalising $\overline{k} = 1$ in (2.31). In view of (2.32), incompressibility

means
$$\nabla \vec{v}_\pm = 0 \quad \text{in} \quad \Omega_\pm(t). \tag{6.2b}$$

The pressure jump across the interface $\Gamma(t)$ is compensated by the surface tension in the interface. From the Laplace-Young condition (2.33) we then obtain

$$u_+ - u_- = \gamma \kappa_{\Gamma(t)} + g(\rho_+ - \rho_-)y \quad \text{on} \quad \Gamma(t), \tag{6.2c}$$

where γ is the surface tension coefficient of the interface, and $\kappa_{\Gamma(t)}$ is the curvature of $\Gamma(t)$. On the bottom Γ_{-1} we impose the no-flux condition

$$\partial_\nu u_- = 0 \quad \text{on} \quad \Gamma_{-1}, \tag{6.2d}$$

with $\nu(t)$ the outward orientated normal at $\partial \Omega_-(t)$, and we assume that the pressure on the boundary Γ_1 is kept constant to be 0

$$u_+ = g\rho_+ \quad \text{on} \quad \Gamma_1. \tag{6.2e}$$

Two more, so called kinematic boundary conditions on $\Gamma(t)$ are given (see also [35, 66, 67]) by

$$V(t,\cdot) = \langle \vec{v}_\pm(t,\cdot), \nu(t,\cdot) \rangle \quad \text{on} \quad \Gamma(t), \tag{6.2f}$$

meaning that the interface moves along with the fluids (a particle on the boundary remains on the boundary as time elapses). Finally, the interface at time $t = 0$ is considered to be known

$$\Gamma(0) = \Gamma_0. \tag{6.2g}$$

The system (6.2) is a two-phase moving boundary problem. The main interest is determining the motion of the interface $\Gamma(t)$ separating the fluids. The potentials $u_\pm$ can be then determined by solving elliptic mixed boundary value problems. We shall prove that if the boundary Γ_0 is sufficiently small, then, locally in time, problem (6.2) possesses a unique classical solution. To that aim, we first rewrite problem (6.2) in a more accessible way by introducing a parametrisation of the, a priori unknown boundary $\Gamma(t)$.

6.2 Parameterising the boundary

In order to avoid the contact angle problem we consider periodic flows only. Let $\alpha \in (0,1)$ be fixed in the remainder of this work. We define the set of

admissible functions to be
$$\mathcal{V} := \{f \in h^{4+\alpha}(\mathbb{S}) : \|f\|_{C(\mathbb{S})} < 1\},$$
where $\mathbb{S}$ is the unit circle. The small Hölder space $h^{m+\beta}(\mathbb{S})$, $m \in \mathbb{N}$ and $\beta \in (0,1)$, is again the completion of $C^\infty(\mathbb{S})$ in $C^{m+\beta}(\mathbb{S})$. Furthermore, we set $\Gamma_k := \mathbb{S} \times \{k\}$ for $k \in \{-1, 0, 1\}$.

To incorporate time, let $T > 0$ be fixed. If a function $f \in C([0,T], \mathcal{V}) \cap C^1([0,T], h^{1+\alpha}(\mathbb{S}))$ describes the evolution of the interface separating fluids, then, at any time $t \in [0,T]$, we have that $\Omega_\pm(t) = \Omega_\pm(f(t))$, where, for $f \in \mathcal{V}$, we defined the following subsets of Ω :
$$\Omega_-(f) := \{(x,y) : -1 < y < f(x)\}, \quad \Omega_+(f) := \{(x,y) : f(x) < y < 1\}.$$
Consequently, it holds that $\Gamma(t) = \Gamma(f(t))$ for all $t \in [0,T]$, where
$$\Gamma(f) := \{(x, f(x)) : x \in \mathbb{S}\}$$
if $f \in \mathcal{V}$. Let us describe the kinematic condition (6.2f) in this particular context. Consider the evolution of a particle $(x(t), y(t))$ on the interface $\Gamma(f(t))$. Since particles on the interface separating the fluids remain there as time evolves, we obtain from the definition of $\Gamma(f)$, that $y(t) = f(t, x(t))$ for all $t \in [0,T]$. Derivation of this equation with respect to the time variable t yields $\partial_t f = (-\partial_x f, 1)(x', y')$, hence, the normal velocity V satisfies the relation
$$V = \frac{\partial_t f}{\sqrt{1 + (\partial_x f)^2}}.$$
Thereby, the local study of (6.2) reduces to the following problem
$$\begin{cases} \Delta u_\pm = 0 & \text{in } \Omega_\pm(f(t)), \; 0 \leq t \leq T, \\ u_+ = \overline{u} & \text{on } \Gamma_1, \; 0 \leq t \leq T, \\ \partial_\nu u_- = 0 & \text{on } \Gamma_{-1}, \; 0 \leq t \leq T, \\ u_+ - u_- = \gamma \kappa_{\Gamma(f)} + \varpi f & \text{on } \Gamma(f(t)), \; 0 \leq t \leq T, \\ \partial_t f + \dfrac{\sqrt{1+(\partial_x f)^2}}{\mu_\pm} \partial_\nu u_\pm = 0 & \text{on } \Gamma(f(t)), \; 0 \leq t \leq T, \\ f(0) = f_0, & \end{cases} \quad (6.3)$$

where $f_0 \in \mathcal{V}$ determines the initial shape of Γ_0, and for simplicity, we set

$$\varpi := g(\rho_+ - \rho_-) \quad \text{and} \quad \bar{u} := g\rho_+. \tag{6.4}$$

Vice versa, defining $\vec{v}_\pm$ by (6.2a), the system (6.2) can be deduced from (6.3). Again, we deal with a volume preserving flow, since:

Observation 6.2.1 (Conservation of volume) *The solutions of the problem* (6.3) *preserve both fluid's quantities.*

Proof Indeed, the cell contains initially a volume $\operatorname{vol}\Omega_-(f(0)) = \int_\mathbb{S} 1 + f(0)\, dx$ of the first fluid. Since

$$\frac{d}{dt}\left(\int_\mathbb{S} 1 + f\, dx\right) = \int_\mathbb{S} \partial_t f\, dx = -\frac{1}{\mu_-}\int_\mathbb{S} \sqrt{1 + (\partial_x f)^2}\partial_\nu u_-\, dx$$

$$= -\frac{2\pi}{\mu_-}\int_{\Gamma_f} \partial_\nu u_-\, ds = -\frac{2\pi}{\mu_-}\int_{\Omega_-(f)} \Delta u_-\, dx$$

$$+ \frac{2\pi}{\mu_-}\int_{\Gamma_{-1}} \partial_\nu u_-\, ds = 0,$$

we conclude that this quantity is preserved by the flow. The volume of the second fluid $\operatorname{vol}\Omega_+(f(t)) = \int_\mathbb{S}(1 - f(t))\, dx$ is also preserved from the same reason. □

This observation serves as motivation for considering in the remainder of this part that the cell contains equal volumes of both fluids. The initial data f_0 of the problem must be taken then from the set $\mathcal{V}_0 := \mathcal{V} \cap h_0^{4+\alpha}(\mathbb{S})$, where

$$h_0^{4+\alpha}(\mathbb{S}) := \{f \in h_0^{4+\alpha}(\mathbb{S}) : \widehat{f}(0) = 0\}.$$

A triple (f, u_+, u_-) is called classical Hölder solution of (6.3) on $[0,T]$ if

$$f \in C([0,T], \mathcal{V}_0) \cap C^1([0,T], h_0^{1+\alpha}(\mathbb{S})),$$

$$u_\pm(t, \cdot) \in \textbf{\textit{buc}}^{2+\alpha}(\Omega_\pm(f(t))), \quad t \in [0,T],$$

and if (f, u_+, u_-) satisfies (6.3) pointwise. Given $f \in \mathcal{V}$, the Banach space $\textbf{\textit{buc}}^{2+\alpha}(\Omega_\pm(f))$ is the completion of $BUC^\infty(\Omega_\pm(f))$ in $BUC^{2+\alpha}(\Omega_\pm(f))$. If we know the mapping f, then the potentials u_+ and u_- can be determined by soving the mixed boundary value problems (7.1) and (7.3), respectively. That is why we shall also say that the function f is the solution of (6.3).

In order to state the first main result of this part we introduce first appropriate Banach spaces. Let $\mathcal{W}$ be an open neighbourhood of the zero function in $h_0^{4+\alpha}(\mathbb{S})$ with the property that $3\mathcal{W} \subset \mathcal{V}_0$. Then

$$\mathcal{U} := \{f \,:\, f \in C([0,T],\mathcal{W}),\, f(0) = 0\} \cap C^1([0,T], h_0^{1+\alpha}(\mathbb{S})),$$

is an open neighbourhood of 0 in the Banach space

$$\mathbb{E} := C_0([0,T], h_0^{4+\alpha}(\mathbb{S})) \cap C^1([0,T], h_0^{1+\alpha}(\mathbb{S})),$$

where

$$C_0\left([0,T], h_0^{4+\alpha}(\mathbb{S})\right) := \left\{f \in C\left([0,T], h_0^{4+\alpha}(\mathbb{S})\right) \,:\, f(0) = 0\right\}.$$

Let $\widehat{\mathbb{E}}$ denote the subspace of $\mathbb{E}$ consisting only of functions f with derivative 0 in $h_0^{1+\alpha}(\mathbb{S})$ at $t = 0$

$$\widehat{\mathbb{E}} := \{f \in \mathbb{E} \,:\, \partial_t f(0) = 0\}.$$

The first main result of this part is the following existence, uniqueness and analytic dependence theorem:

Theorem 6.2.2 (Well-posedness) *Let $T > 0$ be given. There exists an open neighbourhood $\mathcal{O}$ of 0 in $h_0^{4+\alpha}(\mathbb{S})$, such that for all $f_0 \in \mathcal{O}$ problem (6.3) possesses a unique classical Hölder solution $(f = \mathcal{F}(f_0), u_+, u_-)$ on $[0,T]$.*

Moreover, there exist $\delta > 0$, a function $\mathcal{E}(f_0) \in \mathcal{U}$ and a unique mapping $h \in \overline{B}_{\widehat{\mathbb{E}}}(0, 2\delta)$ such that $f = f_0 + \mathcal{E}(f_0) + h$. The solution operator

$$\mathcal{O} \ni f_0 \mapsto \mathcal{F}(f_0) \in C([0,T], \mathcal{V}_0) \cap C^1([0,T], h_0^{1+\alpha}(\mathbb{S}))$$

is analytic.

Observation 6.2.3 *The proof of Theorem 6.2.2 shows that the problem is well-posed also if the surface tension effects are neglected, i.e. $\gamma = 0$, provided the fluid on the bottom of the cell is more dense than that above $\rho_- > \rho_+$. The result of Theorem 6.2.2 remains valid with the obvious modification that $h_0^{4+\alpha}(\mathbb{S})$ is replaced by $h_0^{2+\alpha}(\mathbb{S})$.*

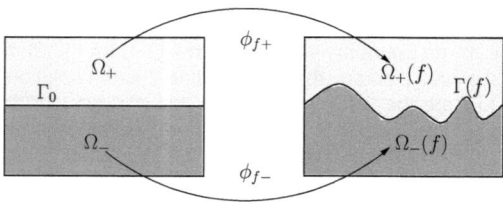

Figure 6.2: The diffeomorphisms $\phi_{f\pm}$.

6.3 The transformed system

In order to prove Theorem 6.2.2 we transform system (6.3) into a problem on a fixed domain. Let us first introduce some notation. We set $\Omega_\pm := \Omega_\pm(0)$. The boundary $\Gamma_0 = \mathbb{S} \times \{0\}$ is also identified with the unit circle $\mathbb{S}$. For every $f \in \mathcal{V}$, we define the mappings $\phi_{f\pm} = (\phi^1_{f\pm}, \phi^2_{f\pm}) : \Omega_\pm \to \Omega_\pm(f)$ by
$$\phi_{f\pm}(x,y) := (x, y + (1 \mp y)f(x)), \quad (x,y) \in \Omega_\pm.$$
The Jacobian of $\phi_{f\pm}$ is given by
$$\partial \phi_{f\pm}(x,y) = \begin{bmatrix} 1 & 0 \\ (1 \mp y)f'(x) & 1 \mp f(x) \end{bmatrix}, \quad (x,y) \in \Omega_\pm.$$

Notice that $\partial\phi_{f\pm}(x,y)$ is invertible for any $(x,y) \in \Omega_\pm$, since $f \in \mathcal{V}$ implies $\|f\|_{C(\mathbb{S})} < 1$. Indeed, given $x \in \mathbb{S}$, we have $\partial_2 \phi^2_{f\pm}(x,y) = 1 \mp f(x) > 0$ for all y. Consequently, $\phi_{f\pm}$ are diffeomorphisms mapping $\Omega_\pm$ onto $\Omega_\pm(f)$, i.e. $\phi_{f\pm} \in \text{Diff}^{4+\alpha}(\Omega_\pm, \Omega_\pm(f))$. Notice that $\phi_{f\pm}(\Gamma_{\pm 1}) = \Gamma_{\pm 1}$ and $\phi_{f\pm}(\Gamma_0) = \Gamma(f)$ (see Figure 6.2). The inverse $\psi_{f\pm}$ of $\phi_{f\pm}$ is given by the expression
$$\psi_{f\pm}(x,y) := \phi^{-1}_{f\pm}(x,y) = \left(x, \frac{y - f(x)}{1 \mp f(x)}\right), \quad (x,y) \in \Omega_\pm(f),$$
and
$$\partial \psi_{f\pm}(\phi_{f\pm}(x,y)) = \begin{bmatrix} 1 & 0 \\ \frac{-(1\mp y)f'(x)}{1 \mp f(x)} & \frac{1}{1 \mp f(x)} \end{bmatrix}, \quad (x,y) \in \Omega_\pm.$$

The push-forward and pull-back operators induced by the diffeomorphisms $\phi_{f\pm}$ are defined by
$$\phi_*^{f\pm} : \text{buc}^{2+\alpha}(\Omega_\pm) \to \text{buc}^{2+\alpha}(\Omega_\pm(f)), \quad v \mapsto v \circ \psi_{f\pm},$$
$$\phi_{f\pm}^* : \text{buc}^{2+\alpha}(\Omega_\pm(f)) \to \text{buc}^{2+\alpha}(\Omega_\pm), \quad u \mapsto u \circ \phi_{f\pm}.$$

It follows by the mean value theorem that $\phi_{f\pm}^*$ and $\phi_*^{f\pm}$ are isomorphisms and that they are inverse to each other $(\phi_{f\pm}^*)^{-1} = \phi_*^{f\pm}$. We use now these isomorphisms to transform system (6.3) on fixed domains. The disadvantage of this method is that the differential operators we have to study are more involved. Given $f \in \mathcal{V}$, we define
$$\mathcal{A}_\pm(f) : \text{buc}^{2+\alpha}(\Omega_\pm) \to \text{buc}^\alpha(\Omega_\pm), \qquad \mathcal{A}\pm(f) := \phi_{f\pm}^* \Delta \phi_*^{f\pm}.$$

The uniformly elliptic operators $\mathcal{A}_\pm(f)$ depend analytically on the a priori unknown function f, i.e.
$$\mathcal{A}_\pm \in C^\omega(\mathcal{V}, \mathcal{L}(\text{buc}^{2+\alpha}(\Omega_\pm), \text{buc}^\alpha(\Omega_\pm))), \tag{6.5}$$

since we have the following representation for $\mathcal{A}_\pm$
$$\mathcal{A}_\pm(f) = \frac{\partial^2}{\partial x^2} - 2\frac{(1 \mp y)f'}{1 \mp f}\frac{\partial^2}{\partial x \partial y} + \left(\frac{(1 \mp y)^2 f'^2}{(1 \mp f)^2} + \frac{1}{(1 \mp f)^2}\right)\frac{\partial^2}{\partial y^2} +$$
$$-(1 \mp y)\frac{(1 \mp f)f'' \pm 2f'^2}{(1 \mp f)^2}\frac{\partial}{\partial y} \quad \text{for } f \in \mathcal{V}.$$

Indeed, $\mathcal{A}_\pm(f)$ is a linear combination of differential operators of order less or equal to 2 with coefficients depending analytically on f. Furthermore, $\mathcal{A}_\pm(f)$ have the following geometric interpretation. Let η be the standard metric on $\mathbb{R}^2$. The diffeomorphisms $\phi_{f\pm}$ induce a Riemannian metric, $\phi_{f\pm}^*\eta$, on $\Omega_\pm$, i.e.
$$\phi_{f\pm}^*\eta\big|_x(\xi,\zeta) := \eta\big|_{\phi_{f\pm}(x)}(\partial\phi_{f\pm}(x)\xi, \partial\phi_{f\pm}(x)\zeta) = \xi^T(\phi_{f\pm}(x))^T \partial\phi_{f\pm}(x)\zeta$$

for $x \in \Omega_\pm$ and tangent vectors $\xi, \zeta \in T_x\Omega_\pm$. With this notation, $\mathcal{A}_\pm(f)$ are exactly the Laplace-Beltrami operator corresponding to the Riemannian manifold $(\Omega_\pm, \phi_{f\pm}^*\eta)$, respectively. Notice also that $\mathcal{A}_\pm$ depends nonlinearly of f.

Lastly, given $f \in \mathcal{V}$, the boundary operators $\mathcal{B}_\pm(f) : \mathsf{buc}^{2+\alpha}(\Omega_\pm) \to h^{1+\alpha}(\mathbb{S})$ are defined by the relation

$$B_\pm(f)v_\pm := \mu_\pm^{-1} \operatorname{tr} \phi_{f\pm}^* \langle \nabla(\phi_*^{f\pm} v_\pm) | (-\partial_x f, 1) \rangle,$$

where tr is the trace operator with respect to $\mathbb{S}$, i.e. $\operatorname{tr} v_\pm(x) := v_\pm(x, 0)$ for all $x \in \mathbb{S}$ and $v_\pm \in BUC(\Omega_\pm)$. A simple computation shows that

$$\mathcal{B}_\pm(f)v_\pm = \frac{1}{\mu_\pm}\left(\frac{1+f'^2}{1 \mp f} \operatorname{tr} \frac{\partial v_\pm}{\partial y} - f' \operatorname{tr} \frac{\partial v_\pm}{\partial x}\right).$$

Obviously, $\mathcal{B}_\pm$ are also real analytic

$$\mathcal{B}_\pm \in C^\omega(\mathcal{V}, \mathcal{L}(\mathsf{buc}^{2+\alpha}(\Omega_\pm), h^{1+\alpha}(\mathbb{S}))). \tag{6.6}$$

Letting $v_\pm = \phi_{f\pm}^* u_\pm$, we see that problem (6.3) is equivalent to the following problem

$$\begin{cases} \mathcal{A}_\pm(f)v_\pm = 0 & \text{in } \Omega_\pm, \quad 0 \le t \le T, \\ v_+ = \overline{u} & \text{on } \Gamma_1, \quad 0 \le t \le T, \\ \partial_\nu v_- = 0 & \text{on } \Gamma_{-1}, \quad 0 \le t \le T, \\ v_+ - v_- = \gamma\kappa(f) + \varpi f & \text{on } \Gamma_0, \quad 0 \le t \le T, \\ \partial_t f + \mathcal{B}_\pm(f)v_\pm = 0 & \text{on } \Gamma_0, \quad 0 \le t \le T, \\ f(0) = f_0, \end{cases} \tag{6.7}$$

where $\kappa : \mathcal{V} \subset h^{4+\alpha}(\mathbb{S}) \to h^{2+\alpha}(\mathbb{S})$, with

$$\kappa(f) = \frac{f''}{(1+f'^2)^{3/2}} \quad \text{for } f \in \mathcal{V},$$

is the pulled-back curvature operator. Obviously, κ depends analytically on f, i.e. $\kappa \in C^\omega(\mathcal{V}, h^{2+\alpha}(\mathbb{S}))$.

A triple (f, v_+, v_-) is called classical Hölder solution of (6.3) on $[0,T]$ if

$$f \in C([0,T], \mathcal{V}_0) \cap C^1([0,T], h_0^{1+\alpha}(\mathbb{S})),$$

$$v_\pm(t, \cdot) \in \mathsf{buc}^{2+\alpha}(\Omega_\pm), \quad t \in [0,T],$$

and if (f, v_+, v_-) satisfies (6.7) pointwise. The systems (6.3) and (6.7) are equivalent in the following sense:

Lemma 6.3.1 *Let (f, u_+, u_-) be a solution of* (6.3). *Then $(f, \phi^*_{f_+} u_+, \phi^*_{f_-} u_-)$ is a solution to* (6.7). *Vice versa, if (f, v_+, v_-) is a classical solution of* (6.7), *then $(f, \phi^{f_+}_* v_+, \phi^{f_-}_* v_-)$ is a classical solution of* (6.3).

Proof The main difficulty lies in showing that, if $f \in \mathcal{V}$, then we have that $u_\pm \in buc^{2+\alpha}(\Omega_\pm(f))$ exactly when $v_\pm \in buc^{2+\alpha}(\Omega_\pm)$. However, this equivalence can be proved by using the mean value theorem as in [27, Lemma 1.2]. □

Although the transformation we made above has the draw-back of introducing additional nonlinear coefficients, it allows us to write the problem as a nonlinear operator equation, as we show in the next chapter. Notice also that the mapping f, describing the moving boundary separating the fluids, is preserved by this transformation.

Chapter 7

The operator equation

In this chapter we shall identify, after constructing a first approximation of the solution in Lemma 7.1.5, the transformed problem (6.7) with an operator equation between some appropriate Banach spaces. By considering the characterisation of the Hölder spaces over the unit circle using the dyadic decomposition, as in Chapter 2, we shall prove in the second section of this chapter the well-posedness result established in Theorem 6.2.2.

7.1 The operator equation

We begin this section by defining solution operators to mixed boundary value problems which are closely related to our system (6.7). By composing this operators we obtain thereafter an operator equation. The solutions of this equation are exactly the solutions of (6.7). Moreover, the operators defined below will reveal that it is crucial to determine the mapping f, describing the evolution of the interface. The potentials $v_\pm$ are then the image of f under these operators.

Given $f \in \mathcal{V}$ and $q \in h^{1+\alpha}(\mathbb{S})$, we let $\mathcal{T}(f,q) \in buc^{2+\alpha}(\Omega_+)$ denote the solution of the linear, elliptic mixed boundary value problem

$$\begin{cases} \mathcal{A}_+(f)v_+ &= 0 \text{ in } \Omega_+, \\ v_+ &= \bar{u} \text{ on } \Gamma_1, \\ \mathcal{B}_+(f)v_+ &= q \text{ on } \Gamma_0, \end{cases} \quad (7.1)$$

with $\bar{u}$ the constant defined by relation (6.4). From (6.5), (6.6), and taking also into consideration that the operator mapping a bijective linear operator onto its inverse is analytical, we get that $\mathcal{T}$ is a real analytic operator.

Lemma 7.1.1 *The operator $\mathcal{T}$ is real analytic, i.e.*
$$\mathcal{T} \in C^\omega(\mathcal{V} \times h^{1+\alpha}(\mathbb{S}), \text{buc}^{2+\alpha}(\Omega_+)).$$
Given $(f_0, q_0) \in \mathcal{V} \times h^{1+\alpha}(\mathbb{S})$, the derivative $\partial\mathcal{T}(f_0, q_0)[f, q]$, with $(f, q) \in h^{4+\alpha}(\mathbb{S}) \times h^{1+\alpha}(\mathbb{S})$, is the solution of the problem
$$\begin{cases} \mathcal{A}(f_0)w_+ = -\partial\mathcal{A}(f_0)[f]\mathcal{T}(f_0, q_0) & \text{in } \Omega_+, \\ w_+ = 0 & \text{on } \Gamma_1, \\ \mathcal{B}_+(f_0)w_+ = q - \partial\mathcal{B}_+(f_0)[f]\mathcal{T}(f_0, q_0) & \text{on } \Gamma_0. \end{cases} \quad (7.2)$$

Proof We are left to determine the Fréchet derivative of $\mathcal{T}$. With $v_+ = \mathcal{T}(f_0 + f, q_0 + q)$, $u_+ := \mathcal{T}(f_0, q_0)$ and $z_+ := v_+ - u_+ - w_+$, we find from the equality
$$\mathcal{A}(f_0 + f)v_+ - \mathcal{A}(f_0)u_+ - \mathcal{A}(f_0)w_+ - \partial\mathcal{A}(f_0)u_+ = 0, \quad \text{in } \Omega_+,$$
that
$$\mathcal{A}(f_0+f)z_+ = -(\mathcal{A}(f_0+f) - \mathcal{A}(f_0))w_+ - (\mathcal{A}(f_0+f) - \mathcal{A}(f_0) - \partial\mathcal{A}(f_0)[f])u_+$$
in Ω_+. The right hand side is of order $O(\|[f,q]\|^2_{C^{4+\alpha}(s) \times C^{1+\alpha}(\mathbb{S})})$ for small (f, q). Furthermore, $z_+ = 0$ on Γ_1, and on Γ_0 we have that
$$0 = \mathcal{B}_+(f_0+f)v_+ - q - q_0 - \mathcal{B}_+(f_0)u_+ + q_0 - \mathcal{B}_+(f_0)w_+ + q - \partial\mathcal{B}(f_0)[f]u_+,$$
hence
$$\mathcal{B}_+(f_0+f)z_+$$
$$= -(\mathcal{B}_+(f_0+f) - \mathcal{B}_+(f_0))w_+ - (\mathcal{B}_+(f_0+f) - \mathcal{B}_+(f_0) - \partial\mathcal{B}_+(f_0)[f])u_+,$$
and the right hand side is also of order $O(\|[f,q]\|^2_{C^{4+\alpha}(s) \times C^{1+\alpha}(\mathbb{S})})$. We conclude, using estimates for mixed boundary value problem presented in Theorem 7.3 and Remark 2 on page 669 in [1], that $\partial\mathcal{T}(f_0, q_0)[h, q]$ is the solution of (7.2). This completes the proof.

Furthermore, we define the operator $\mathcal{S}: \mathcal{V}\times buc^{2+\alpha}(\Omega_+) \to buc^{2+\alpha}(\Omega_-)$ by writing $\mathcal{S}(f, v_+)$ for the unique solution of the problem

$$\begin{cases} \mathcal{A}_-(f)v_- = 0 & \text{in } \Omega_-, \\ v_- = v_+ - \gamma\kappa(f) - \varpi f & \text{on } \Gamma_0, \\ \partial_\nu v_- = 0 & \text{on } \Gamma_{-1}. \end{cases} \quad (7.3)$$

Let us notice that $\mathcal{S}$ depends analytically on its variables as well. Similarly to Lemma 7.1.1, we obtain:

Lemma 7.1.2 *Given $(f_0, v_{+_0}) \in \mathcal{V} \times buc^{2+\alpha}(\Omega_+)$ and $(f, v_+) \in h^{4+\alpha}(\mathbb{S}) \times buc^{2+\alpha}(\Omega_+)$, the Fréchet derivative $\partial \mathcal{S}(f_0, v_{+0})[f, v_+]$ is the solution of the problem*

$$\begin{cases} \mathcal{A}(f_0)w_- = -\partial\mathcal{A}(f_0)[f]\mathcal{S}(f_0, v_{+0}) & \text{in } \Omega_+, \\ w_- = v_+ - \gamma\partial\kappa(f_0)[f] - \varpi f & \text{on } \Gamma_0, \\ \partial_\nu w_- = 0 & \text{on } \Gamma_{-1}. \end{cases} \quad (7.4)$$

Proof The proof is similar to that of Lemma 7.1.1 and we omit it. □

With these definitions we set that if (f, v_+, v_-) is a classical Hölder solution of (6.7), then $v_+ = \mathcal{T}(f, -\partial_t f)$, $v_- = \mathcal{S}(f, \mathcal{T}(f, -\partial_t f))$, and

$$\partial_t f + \mathcal{B}_-(f)\mathcal{S}(f, \mathcal{T}(f, -\partial_t f)) = 0 \quad \text{in } [0, T]. \quad (7.5)$$

Conversely, if $f \in C([0, T], \mathcal{V}_0) \cap C^1([0, T], h_0^{1+\alpha}(\mathbb{S}))$ is a solution of (7.5) satisfying initially $f(0) = f_0$, then (f, v_+, v_-), with v_+ and v_- defined above, is a classical Hölder solution of (6.7). Hence, the study of (6.7) reduces to the study of the operator equation (7.5).

Let us now consider the operator on the left hand side of equation (7.5) and study its properties. We show that for all $(f, q) \in \mathcal{V} \times h^{1+\alpha}(\mathbb{S})$, the function $\mathcal{B}_-(f)(f,\mathcal{S}(f,\mathcal{T}(f,q)))$ has integral mean 0. Indeed, let $v_+ := \mathcal{T}(f,q)$ and $v_- := \mathcal{S}(f,v_+)$. Using Stokes' theorem, we compute

$$\mu_- \int_{\mathbb{S}} \mathcal{B}_-(f)\mathcal{S}(f,\mathcal{T}(f,q))\,dx$$

$$= \int_{\mathbb{S}} \left(\frac{1+f'^2(x)}{1+f(x)} \frac{\partial v_-}{\partial y}(x,0) - f'(x)\frac{\partial v_-}{\partial x}(x,0) \right) dx$$

$$= \int_{\mathbb{S}} (-f'(x), 1) \cdot \nabla(\phi_*^{f^-} v_-)(x,f(x))\,dx = \frac{1}{2\pi} \int_{\Gamma(f)} \partial_\nu \left(\phi_*^{f^-} v_- \right) ds$$

$$= \frac{1}{2\pi} \int_{\Omega_-(f)} \Delta \left(\phi_*^{f^-} v_- \right) dx - \frac{1}{2\pi} \int_{\Gamma_{-1}} \partial_\nu \left(\phi_*^{f^-} v_- \right) ds = 0,$$

since $\phi_*^{f^-} v_-$ is the solution of the mixed boundary value problem

$$\begin{cases} \Delta u_- = 0 & \text{in } \Omega_-(f), \\ u_- = \phi_*^{f^+} v_+ - \gamma\kappa_{\Gamma(f)} - \varpi f & \text{on } \Gamma(f), \\ \partial_\nu u_- = 0 & \text{on } \Gamma_{-1}. \end{cases}$$

Hence, if $f \in C([0,T],\mathcal{V}_0) \cap C^1([0,T],h_0^{1+\alpha}(\mathbb{S}))$, then

$$\partial_t f + \mathcal{B}_-(f)\mathcal{S}(f,\mathcal{T}(f,-\partial_t f)) \in \mathbb{F} := C([0,T],h_0^{1+\alpha}(\mathbb{S})).$$

Finally, we define the mapping $G : \mathcal{W} \times \mathcal{U} \subset h_0^{4+\alpha}(\mathbb{S}) \times \mathbb{E} \to \mathbb{F}$, by the relation

$$G(f,h) = \partial_t h + \mathcal{B}_-(f+h)\mathcal{S}(f+h,\mathcal{T}(f+h,-\partial_t h)). \tag{7.6}$$

Our goal is to determine, for each f, a function $h \in \mathcal{U}$ such that

$$G(f,h) = 0. \tag{7.7}$$

Then, $\mathcal{F}(f) := f + h$ is a solution to our problem (6.7), with initial data $\mathcal{F}(f)(0) = f + h(0) = f$. We shall solve this equation using Newton's iteration method (cf. [20, Theorem 15.6]).

Theorem 7.1.3 (Newton's iteration method) *Let X and Y be two Banach spaces, and $F : B_X(0,r) \subset X \to Y$ a C^1–map such that:*

(a) $F'(0)^{-1} \in \mathcal{L}is(Y,X)$, $\|F'(0)^{-1}F(0)\|_X \leq \delta$, and $\|F'(0)^{-1}\|_{\mathcal{L}(Y,X)} \leq \beta$;

(b) $\|F'(x) - F'(\overline{x})\|_{\mathcal{L}(X,Y)} \leq k\|x - \overline{x}\|_X$ for all $x, \overline{x} \in B_X(0,r)$;

(c) $2k\delta\beta < 1$ and $2\delta < 1$,

are satisfied. Then F has a unique zero z in $\overline{B}_X(0, 2\delta)$, and the Newton iterates

$$x_{n+1} = x_n - F'(x_n)^{-1}F(x_n)$$

converge quadratically to z.

A direct application of [20, Theorem 15.6] to the operator equation (7.7) seems very natural. However, if we fix $f \in \mathcal{W}$ and compute the derivative $\partial_h G(f, 0)$ it is not possible to show that $\partial_h G(f, 0)$ is an isomorphism, even when $f = 0$. Therefore, we construct first an approximation of the solution of (6.3), which will make it possible for us to write (7.7) as an operator equation between subspaces $\widehat{\mathbb{E}}$ and $\widehat{\mathbb{F}}$ of $\mathbb{E}$ and $\mathbb{F}$, respectively. The functions in $\widehat{\mathbb{E}}$ and $\widehat{\mathbb{F}}$ have special properties which shall enable us to prove that the assumptions of Theorme 7.1.3 are fulfilled by the operator acting between them.

Let us first observe that if $(f, h) \in \mathcal{W} \times \mathcal{U}$ satisfies $G(f, h) = 0$, meaning that $\mathcal{F}(f) = f + h$ is a solution of (6.3) corresponding to the initial data f then, at time $t = 0$, we have that $\partial_t \mathcal{F}(f)(0) = \partial_t h(0)$ solves

$$\partial_t h(0) + \mathcal{B}_-(f)\mathcal{S}(f, \mathcal{T}(f, -\partial_t h(0))) = 0.$$

The well-posedness of problem (6.3) would imply that, for each f in some small neighbourhood $\mathcal{O}$ of 0 in $h_0^{4+\alpha}(\mathbb{S})$, there exists at least a function $g := \partial_t \mathcal{F}(f)(0) \in h_0^{1+\alpha}(\mathbb{S})$ such that

$$g + \mathcal{B}_-(f)\mathcal{S}(f, \mathcal{T}(f, -g)) = 0. \tag{7.8}$$

Indeed, using the implicit function theorem we get:

Lemma 7.1.4 *There exists an open neighbourhood $\mathcal{O} \subset \mathcal{W}$ of 0 in $h_0^{4+\alpha}(\mathbb{S})$ with the property that for all $f \in \mathcal{O}$ there is a unique function $g := \mathcal{I}(f)$ in $h_0^{1+\alpha}(\mathbb{S})$, such that*

$$\mathcal{G}(f, g) := g + \mathcal{B}_-(f)\mathcal{S}(f, \mathcal{T}(f, -g)) = 0.$$

Moreover, the mapping

$$\mathcal{O} \ni f \mapsto \mathcal{I}(f) \in h_0^{1+\alpha}(\mathbb{S})$$

is analytic.

Proof Assume that we can write

$$\mathcal{G}(f,g) = \mathcal{G}_1(f)g + \mathcal{G}_2(f) \tag{7.9}$$

where $\mathcal{G}_1 \in C^\omega(\mathcal{V}_0, \mathcal{L}(h_0^{1+\alpha}(\mathbb{S}), h_0^{1+\alpha}(\mathbb{S})))$ and $\mathcal{G}_2 \in C^\omega(\mathcal{V}_0, h_0^{1+\alpha}(\mathbb{S}))$. Moreover, let us presuppose that $\mathcal{G}_1(0) \in \mathcal{L}is(h_0^{1+\alpha}(\mathbb{S}))$. From the continuity of $\mathcal{G}_1$ we find then an open neighbourhood $\mathcal{O}$ of 0 in $h_0^{1+\alpha}(\mathbb{S})$ such that $\mathcal{G}_1(f) \in \mathcal{L}is(h_0^{1+\alpha}(\mathbb{S}))$ for all $f \in \mathcal{O}$. Consequently, $\mathcal{I}(f) := -(\mathcal{G}_1(f))^{-1}\mathcal{G}_2(f)$ is the unique solution of the equation $\mathcal{G}(f,\cdot) = 0$ in $h_0^{1+\alpha}(\mathbb{S})$. Furthermore, the mapping $\mathcal{I}$ is real analytic.

So, let us check first that relation (7.9) is true. In view of (7.1), we may decompose $\mathcal{T}(f,g) = \bar{u} + \mathcal{T}_1(f)g$, $f \in \mathcal{V}_0$ and $g \in h_0^{1+\alpha}(\mathbb{S})$, where $\mathcal{T}_1(f)g$ is the solution of the mixed boundary problem

$$\begin{cases} \mathcal{A}_+(f)v_+ = 0 & \text{in } \Omega_+, \\ v_+ = 0 & \text{on } \Gamma_1, \\ \mathcal{B}_+(f)v_+ = g & \text{on } \Gamma_0. \end{cases} \tag{7.10}$$

Similarly, we write $\mathcal{S}(f, v_+) = \mathcal{S}_1(f)v_+ + \mathcal{S}_2(f)$, $f \in \mathcal{V}_0$ and $v_+ \in buc^{2+\alpha}(\Omega_+)$, where $\mathcal{S}_1(f)v_+$ is the solution of

$$\begin{cases} \mathcal{A}_-(f)v_- = 0 & \text{in } \Omega_-, \\ v_- = v_+ & \text{on } \Gamma_0, \\ \partial_\nu v_- = 0 & \text{on } \Gamma_{-1}, \end{cases} \tag{7.11}$$

and $\mathcal{S}_2(f)$ solves

$$\begin{cases} \mathcal{A}_-(f)v_- = 0 & \text{in } \Omega_-, \\ v_- = -\gamma\kappa(f) - \varpi f & \text{on } \Gamma_0, \\ \partial_\nu v_- = 0 & \text{on } \Gamma_{-1}. \end{cases}$$

Notice that $\mathcal{T}_1(f)$ and $\mathcal{S}_1(f)$ are bounded linear operators for all $f \in \mathcal{V}_0$. Whence, we then have

$$\mathcal{G}(f,g) = g + \mathcal{B}_-(f)\mathcal{S}(f, \mathcal{T}(f,-g)) = g + \mathcal{B}_-(f)\mathcal{S}_1(f)\mathcal{T}(f,-g) + \mathcal{B}_-(f)\mathcal{S}_2(f)$$
$$= g - \mathcal{B}_-(f)\mathcal{S}_1(f)\mathcal{T}_1(f)g + \mathcal{B}_-(f)\mathcal{S}_2(f),$$

and setting $\mathcal{G}_1(f)g := g - \mathcal{B}_-(f)\mathcal{S}_1(f)\mathcal{T}_1(f)g$ and $\mathcal{G}_2(f) := \mathcal{B}_-(f)\mathcal{S}_2(f)$, we obtain the relation (7.9). In order to complete the proof, we are left to check that $\mathcal{G}_1(0) \in \mathcal{L}is(h_0^{1+\alpha}(\mathbb{S}))$. We pick $g \in h_0^{1+\alpha}(\mathbb{S})$ and consider its Fourier expansion $g = \sum_{k \in \mathbb{Z}\setminus\{0\}} a_k e^{ikx}$. As in [27], we obtain that v_+, the solution of (7.10) for $f = 0$, has the following expansion

$$v_+(x,y) = \sum_{k \in \mathbb{Z}\setminus\{0\}} \mu_+ \left(\frac{e^{-k}}{k(e^k + e^{-k})} e^{ky} - \frac{e^k}{k(e^k + e^{-k})} e^{-ky} \right) a_k e^{ikx}.$$

Considering the Fourier expansion of $\operatorname{tr} v_+$

$$v_+(x,0) = -\sum_{k \in \mathbb{Z}\setminus\{0\}} \mu_+ \frac{\tanh(k)}{k} a_k e^{ikx},$$

we get that $v_- := \mathcal{S}_1(0)v_+$ has the following expansion

$$v_- = -\sum_{k \in \mathbb{Z}\setminus\{0\}} \mu_+ \left(\frac{e^k}{e^k + e^{-k}} e^{ky} + \frac{e^{-k}}{e^k + e^{-k}} e^{-ky} \right) \frac{\tanh(k)}{k} a_k e^{ikx}.$$

Hence, we determined that $\mathcal{G}_1(0)$ is the Fourier multiplication operator

$$\mathcal{G}_1(0)\left[\sum_{k \in \mathbb{Z}\setminus\{0\}} a_k e^{ikx} \right] = \sum_{k \in \mathbb{Z}\setminus\{0\}} \left(1 + \frac{\mu_+}{\mu_-} \tanh^2(k) \right) a_k e^{ikx}.$$

We notice that the symbol of this operator is bounded away from zero. The Fourier multiplication operator with symbol $(\mu_-/(\mu_- + \mu_+ \tanh^2(k)))_{k \in \mathbb{Z}}$, belongs, in view of Theorem 2.2.1, to $\mathcal{L}(h^{1+\alpha}(\mathbb{S}))$, and its restriction to $h_0^{1+\alpha}(\mathbb{S})$ is the inverse of $\mathcal{G}_1(0)$. We conclude that $\mathcal{G}_1(0)$ is an isomorphism and the proof is complete. $\square$

In the next lemma we construct for each $f \in \mathcal{O}$, a function $\mathcal{E}(f) \in \mathcal{U}$ with the property that $\partial_t \mathcal{E}(f)(0) = \mathcal{I}(f)$. This extension will allow us to re-express equation (7.7) as an operator equation between appropriate subspaces of $\mathbb{E}$ and $\mathbb{F}$. The proof of the lemma relies strongly on the interpolation equation (2.42) and on the maximal regularity theory for the small Hölder spaces. It justifies also why we chose to work with the small Hölder spaces instead of the usual Hölder spaces.

Lemma 7.1.5 *Given $f \in \mathcal{O}$, there exists a function $g := \mathcal{E}(f) \in \mathcal{U}$ with the property that $\partial_t g(0) = \mathcal{I}(f)$. Moreover, the mapping*

$$\mathcal{O} \ni f \mapsto \mathcal{E}(f) \in \mathbb{E}$$

is real analytic.

Proof Let $f \in \mathcal{O}$ be given and consider the inhomogeneous Cauchy problem

$$\begin{cases} x'(t) = Bx(t) + \mathcal{I}(f), & 0 \leq t \leq T, \\ x(0) = 0. \end{cases}$$

We choose B to be the generator of a strongly continuous and analytic semigroup in $\mathcal{L}(h_0^{1+\alpha}(\mathbb{S}))$, i.e. $-B \in \mathcal{H}(h_0^{4+\alpha}(\mathbb{S}), h_0^{1+\alpha}(\mathbb{S}))$. We refer here to [26] where such an operator is studied, or to the operator defined by (4.11). For example we may take

$$B \left[\sum_{k \in \mathbb{Z} \setminus \{0\}} a_k e^{ikx} \right] = - \sum_{k \in \mathbb{Z} \setminus \{0\}} |k|^3 a_k e^{ikx}$$

for $h = \sum_{k \in \mathbb{Z} \setminus \{0\}} a_k e^{ikx} \in h_0^{4+\alpha}(\mathbb{S})$. In fact, $-B \in \mathcal{H}(h_0^{4+\beta}(\mathbb{S}), h_0^{1+\beta}(\mathbb{S}))$ for all $0 < \beta < 1$. For such operators, cf. [51, Corollary 4.3.10], the Cauchy problem is solvable within the set $C^1([0,T], h_0^{4+\alpha}(\mathbb{S})) \cap C([0,T], h_0^{1+\alpha}(\mathbb{S}))$. The function $\mathcal{E}(f) := x$ satisfies then $\mathcal{E}(f)(0) = x(0) = 0$, hence $\mathcal{E}(f) \in \mathbb{E}$, and setting $t = 0$ in the first equation of the Cauchy problem we also get $\partial_t \mathcal{E}(f)(0) = \mathcal{I}(f)$.

Concerning the regularity assumption, standard estimates for the inhomogeneous Cauchy problem imply that

$$\|\mathcal{E}(f)\|_{\mathbb{E}} \leq C \|\mathcal{I}(f)\|_{h_0^{1+\alpha}(\mathbb{S})}, \tag{7.12}$$

where C does not depend on f. By choosing $\mathcal{O}$ small enough in Lemma 7.1.4, we get $\mathcal{E}(f) \in \mathcal{U}$. Moreover, since $\mathcal{I}$ is analytical, we obtain the desired result. $\square$

We transform now the equation (7.7) into an operator equation such that the Theorem 7.1.3 may be applied. We consider first the subspace

$$\widehat{\mathbb{F}} := \{f \in \mathbb{F} : f(0) = 0\},$$

and set $\widehat{\mathcal{U}} := \mathcal{U} \cap \widehat{\mathbb{E}}$. The operator $F : \mathcal{O} \times \widehat{\mathcal{U}} \subset h_0^{4+\alpha}(\mathbb{S}) \times \widehat{\mathbb{E}} \to \widehat{\mathbb{F}}$ is defined by

$$F(f, h) = \partial_t h + \partial_t \mathcal{E}(f)$$
$$+ \mathcal{B}_-(f + \mathcal{E}(f) + h)\mathcal{S}(f + \mathcal{E}(f) + h, \mathcal{T}(f + \mathcal{E}(f) + h, -\partial_t h - \partial_t \mathcal{E}(f))). \tag{7.13}$$

Let us first verify that F is well-defined. In view of $f + \mathcal{E}(f)(t) + h(t) \in 3\mathcal{W} \subset \mathcal{V}_0$ for all $t \in [0, T]$ we must only show that $F(f, h)(0) = 0$. Here comes into play the mapping $\mathcal{E}(f)$ constructed in Lemma 7.1.5. Due to $h(0) = \mathcal{E}(f)(0) = 0$, $\partial_t h(0) = 0$ and $\partial_t \mathcal{E}(f)(0) = \mathcal{I}(f)$, we obtain that

$$F(f, h)(0) = \mathcal{I}(f) + \mathcal{B}_-(f)\mathcal{S}(f, \mathcal{T}(f, -\mathcal{I}(f))),$$

and we infer from Lemma 7.1.4 that $F(f, h)(0) = 0$. Hence, F is well-defined. Combining Lemmas 7.1.1-7.1.2 and relation (6.6), we get that F is smooth in $\mathcal{O} \times \widehat{\mathcal{U}}$ and real analytic in a small neighbourhood of $(0, 0) \in h_0^{4+\alpha}(\mathbb{S}) \times \widehat{\mathbb{E}}$. In the next section we show that if $\mathcal{O}$ is small enough and $f \in \mathcal{O}$, then $F(f, \cdot)$ satisfies the assumptions of Theorem 7.1.3. The function $\mathcal{F}(f) := f + \mathcal{E}(f) + h$, where $h \in \widehat{\mathcal{U}}$ is the unique solution of $F(f, h) = 0$ in a small ball in $\widehat{\mathbb{E}}$ centred in 0, is then the solution of (6.3).

7.2 Proof of Theorem 6.2.2

Let us first notice that, $(f, u_+, u_-) = (0, \overline{u}, \overline{u})$ is a stationary solution of (6.3), hence, the function F defined by (7.13), satisfies $F(0, 0) = 0$. It is well known that the set of isomorphisms is open in $\mathcal{L}(\widehat{\mathbb{E}}, \widehat{\mathbb{F}})$. Consequently, using the smoothness of F, we find out that the assumptions of Theorem 7.1.3 are satisfied, for every f in a small neighbourhood of 0, if we show that $\partial_h F(0,0) \in \mathcal{L}is(\widehat{\mathbb{E}}, \widehat{\mathbb{F}})$.

We begin and determine the derivative $\partial_h F(0,0)$. Plugging $\mathcal{E}(0) = 0$ in (7.13), yields that

$$F(0,h) = \partial_t h + \mathcal{B}_-(h)\mathcal{S}(h, \mathcal{T}(h, -\partial_t h)) \quad \text{for all} \quad h \in \widehat{\mathcal{U}}.$$

By the chain rule we then have

$$\partial_h F(0,0)[h] = \partial_t h + \mathcal{B}_-(0)\partial\mathcal{S}(0, \overline{u})[h, \partial\mathcal{T}(0,0)[h, -\partial_t h]] \quad (7.14)$$

for all $h \in \widehat{\mathbb{E}}$. We made use of the fact that $\mathcal{S}(0, \overline{u}) = \overline{u}$, which simplifies the formula above when $f = 0$, since $\partial \mathcal{B}_-(0)[h]\mathcal{S}(0,\overline{u}) = 0$ for all $h \in \widehat{\mathbb{E}}$. In the following we consider Fourier expansions of the functions $h \in \widehat{\mathbb{E}}$, i.e.

$$h(t, x) = \sum_{k \in \mathbb{Z} \setminus \{0\}} a_k(t) e^{ikx},$$

and aim to determine the Fourier expansion of $\partial_h F(0,0)[h]$. To this scope we shall determine Fourier expansions corresponding to each of the operators involved in relation (7.14). Since h is differentiable with respect to the time variable t, then so are its Fourier coefficients a_k, $k \in \mathbb{Z} \setminus \{0\}$.

Lemma 7.2.1 *Given* $h = \sum_{k \in \mathbb{Z} \setminus \{0\}} a_k(t) e^{ikx} \in \widehat{\mathbb{E}}$, *the derivative* $\partial_h F(0,0)[h]$ *has the following Fourier expansion*

$$\partial_h F(0,0)[h] = \sum_{k \in \mathbb{Z} \setminus \{0\}} \left[\left(1 + \frac{\mu_+}{\mu_-} \tanh^2(k)\right) a'_k + \frac{(\gamma k^3 - \varpi k)\tanh(k)}{\mu_-} a_k \right] e^{ikx}. \quad (7.15)$$

Proof Let $h = \sum_{k \in \mathbb{Z} \setminus \{0\}} a_k(t) e^{ikx} \in \widehat{\mathbb{E}}$ be given. We infer from (7.2) that the derivative $w_+ := \partial \mathcal{T}(0,0)[h, -\partial_t h] \in C([0,T], \text{buc}^{2+\alpha}(\Omega_+))$ is the solution of the following problem

$$\begin{cases} \Delta w_+ = 0 & \text{in } \Omega_+, \ 0 \leq t \leq T, \\ w_+ = 0 & \text{on } \Gamma_1, \ 0 \leq t \leq T, \\ \partial_\nu w_+ = -\mu_+ \partial_t h & \text{on } \Gamma_0, \ 0 \leq t \leq T. \end{cases} \quad (7.16)$$

Considering the following expansion of w_+,

$$w_+(t,x,y) = \sum_{k \in \mathbb{Z}} A_k(t,y) e^{ikx},$$

we find, after plugging it in (7.16) and identifying the coefficients of e^{ikx}, that $A_0 = 0$, and

$$\begin{cases} \partial_y^2 A_k(t,\cdot) - k^2 A_k(t,\cdot) = 0, & \text{in } 0 \le y \le 1, \\ A_k(t,1) = 0, \\ \partial_y A_k(t,0) = -\mu_+ a_k'(t) \end{cases} \tag{7.17}$$

for all $0 \le t \le T$ and $k \in \mathbb{Z} \setminus \{0\}$. The solution of (7.17) is the function

$$A_k(t,y) = -\frac{\mu_+ a_k'(t) e^{-k}}{k(e^k + e^{-k})} e^{ky} + \frac{\mu_+ a_k'(t) e^k}{k(e^k + e^{-k})} e^{-ky}, \quad (t,y) \in [0,T] \times [0,1].$$

By (7.4), the derivative $w_- := \partial S(0,\overline{u})[h,w_+] \in C([0,T], buc^{2+\alpha}(\Omega_-))$ is the solution of the mixed boundary value problem

$$\begin{cases} \Delta w_- = 0 & \text{in } \Omega_+, \quad 0 \le t \le T, \\ w_- = w_+ - \gamma \partial_x^2 h - \varpi h & \text{on } \Gamma_0, \quad 0 \le t \le T, \\ \partial_\nu w_- = 0 & \text{on } \Gamma_{-1}, \quad 0 \le t \le T. \end{cases} \tag{7.18}$$

To obtain (7.18) one has to know that $\partial \kappa(0)[f] = f''$ for all $f \in h_0^{4+\alpha}(\mathbb{S})$. This relation is an immediate consequence of the formula defining $\kappa(f)$. As in the previous step, we expand

$$w_-(t,x,y) = \sum_{k \in \mathbb{Z}} B_k(t,y) e^{ikx}, \quad (t,y) \in [0,T] \times [-1,0],$$

and get that $B_0 = 0$, and

$$\begin{cases} \partial_y^2 B_k(t,\cdot) - k^2 B_k(t,\cdot) = 0, & \text{in } -1 \le y \le 0, \\ B_k(t,0) = \mu_+ \dfrac{\tanh(k)}{k} a_k'(t) + (\gamma k^2 - \varpi) a_k(t), \\ \partial_y B_k(t,-1) = 0 \end{cases}$$

$$\tag{7.19}$$

for all $t \in [0,T]$ and $k \in \mathbb{Z} \setminus \{0\}$. The solution to (7.19) is the function

$$B_k(t,y) = \left(\mu_+ \frac{\tanh(k)}{k} a'_k(t) + (\gamma k^2 - \varpi) a_k(t)\right) \frac{e^k}{e^k + e^{-k}} e^{ky}$$

$$+ \left(\mu_+ \frac{\tanh(k)}{k} a'_k(t) + (\gamma k^2 - \varpi) a_k(t)\right) \frac{e^{-k}}{e^k + e^{-k}} e^{-ky}$$

for $(t,y) \in [0,T] \times [-1,0]$. Summarising, we infer from relation (7.14) that $\partial_h F(0,0)[h] = \partial_t h + \partial_y w_- / \mu_-$, and the expansion (7.15) follows at once. $\square$

Let now $f \in \widehat{\mathbb{F}}$ be given and consider its Fourier expansion

$$f(t,x) = \sum_{k \in \mathbb{Z} \setminus \{0\}} f_k(t) e^{ikx}.$$

If $\partial_h F(0,0)[h] = f$, for some $h = \sum_{k \in \mathbb{Z}\setminus\{0\}} a_k(t) e^{ikx} \in \widehat{\mathbb{E}}$, then a_k is the solution of the following initial value problem

$$\begin{cases} \left(1 + \frac{\mu_+}{\mu_-} \tanh^2(k)\right) a'_k(t) + \frac{\gamma k^3 - \varpi k}{\mu_-} \tanh(k) a_k(t) = f_k(t), & t \in [0,T], \\ a_k(0) = 0. \end{cases}$$
(7.20)

The solution to (7.20) is

$$a_k(t) = \int_0^t K(k, t-s) f_k(s)\, ds, \quad \text{where} \tag{7.21}$$

$$K(\xi, t) := \frac{\mu_-}{\mu_- + \mu_+ \tanh^2(\xi)} e^{-m(\xi)t}, \quad (\xi, t) \in \mathbb{R} \times [0,T], \quad \text{and} \tag{7.22}$$

$$m(\xi) := \frac{\tanh(\xi)}{\mu_- + \mu_+ \tanh^2(\xi)} (\gamma \xi^3 - \varpi \xi), \quad \xi \in \mathbb{R}. \tag{7.23}$$

Let us first notice that if $\partial_h F(0,0)[h] = 0$, then from (7.21), we find that $a_k = 0$ for all $k \in \mathbb{Z} \setminus \{0\}$. We deduce that $h = 0$, hence $\partial_h F(0,0)$ is one-to-one. We are left to show that the operator

$$\widehat{\mathbb{F}} \ni \sum_{k\in\mathbb{Z}\setminus\{0\}} f_k(t)e^{ikx} \xmapsto{A} \sum_{k\in\mathbb{Z}\setminus\{0\}} a_k(t)e^{ikx} \in \widehat{\mathbb{E}}, \qquad (7.24)$$

where a_k, $k \in \mathbb{Z} \setminus \{0\}$, are the functions defined by relation (7.21), is well-defined. This implies that $\partial_h F(0,0)$ is onto and its inverse is exactly $A \in \mathcal{L}(\widehat{\mathbb{F}}, \widehat{\mathbb{E}})$. Obviously, $Af(0) = 0$, and, from the first equation of (7.20), we have that $\partial_t (Af)(0) = 0$ for all $f \in \widehat{\mathbb{F}}$. Theorem 2.2.1 and the first equality in (7.20) reduce the problem of showing that $Af \in \widehat{\mathbb{E}}$ to showing that

$$Af \in C_0([0,T], h_0^{4+\alpha}(\mathbb{S})). \qquad (7.25)$$

Indeed, we have:

Lemma 7.2.2 *Assume that* $Af \in C_0([0,T], h_0^{4+\alpha}(\mathbb{S}))$. *Then* $Af \in \widehat{\mathbb{E}}$.

Proof Similar arguments to those presented in the proof of Lemma 4.2.2 imply, in view of Theorem 2.2.1, that the Fourier multiplication operator

$$\sum_{k\in\mathbb{Z}\setminus\{0\}} a_k e^{ikx} \xmapsto{\mathfrak{M}} \sum_{k\in\mathbb{Z}\setminus\{0\}} \frac{\overline{k}\tanh(k)}{\mu_- + \mu_+ \tanh^2(k)}(\gamma k^3 - \varpi k) a_k e^{ikx}$$

belongs to $\mathcal{L}(h_0^{4+\alpha}(\mathbb{S}), h_0^{1+\alpha}(\mathbb{S}))$. Consequently, the composition $\mathfrak{M} \circ Af$ belongs to $C([0,T], h_0^{1+\alpha}(\mathbb{S}))$. Similarly,

$$h_0^{1+\alpha}(\mathbb{S}) \ni \sum_{k\in\mathbb{Z}\setminus\{0\}} f_k e^{ikx} \xmapsto{\mathfrak{N}} \sum_{k\in\mathbb{Z}\setminus\{0\}} \frac{\mu_-}{\mu_- + \mu_+ \tanh^2(k)} f_k e^{ikx} \in h_0^{1+\alpha}(\mathbb{S})$$

is a continuous mapping. From the first equation of (7.20) we find out that the function

$$[0,T] \ni t \xmapsto{g} \sum_{k\in\mathbb{Z}\setminus\{0\}} a'_k(t)e^{ikx} \in h_0^{1+\alpha}(\mathbb{S})$$

satisfies $g = \mathfrak{M}Af + \mathfrak{N}f$. Consequently, g belongs to $C([0,T], h_0^{1+\alpha}(\mathbb{S}))$, and additionally $g(0) = 0$. It follows then immediately that $Af(t) = \int_0^t g(s)\,ds$ for all $t \in [0,T]$, hence $Af \in \widehat{\mathbb{E}}$. $\qquad\square$

We come now to the crucial point of this section, that of showing (7.25). Given $f \in \widehat{\mathbb{F}}$, we recall once more that $Af(0) = 0$. Hence, we have only to prove that Af is well-defined and continuous with values in $h_0^{4+\alpha}(\mathbb{S})$. The next theorem is obtained by using some of the arguments presented in [67, Lemma 2.2].

Theorem 7.2.3 *Given $f \in \widehat{\mathbb{F}}$, with*

$$f(t,x) = \sum_{k \in \mathbb{Z} \setminus \{0\}} f_k(t) e^{ikx},$$

the function

$$[0,T] \ni t \mapsto Af(t) = \sum_{k \in \mathbb{Z} \setminus \{0\}} a_k(t) e^{ikx} \in h_0^{4+\alpha}(\mathbb{S}),$$

where $a_k, k \in \mathbb{Z} \setminus \{0\}$, are defined by (7.21)-(7.23), is well-defined and continuous.

Proof To make the proof more accessible to the reader we divide it into three steps.

Step 1 We show first that Af is a bounded $Af \in B([0,T], C_0^{4+\alpha}(\mathbb{S}))$ even if $f \in \mathbb{F}$. More precisely, we find a positive constant L (which depends on T) such that

$$\sup_{t \in [0,T]} \|Af(t)\|_{C^{4+\alpha}(\mathbb{S})} \leq L \|f\|_{\mathbb{F}} \quad \text{for all } f \in \mathbb{F}. \tag{7.26}$$

Given $f \in \mathbb{F}$, we find, thanks to the embedding $h_0^{1+\alpha}(\mathbb{S}) \hookrightarrow L_2(\mathbb{S})$, a uniform bound for the Fourier coefficients $(f_k(t))$, i.e. $|f_k(t)| \leq M$ for all $k \in \mathbb{Z}$ and $t \in [0,T]$. The mapping K is bounded on $\mathbb{R} \times [0,T]$, hence $(a_k(t))_k$ is bounded uniformly in $t \in [0,T]$. Consequently, $f(t) \in \mathcal{D}'(\mathbb{S}, \mathbb{C})$ for all $t \in [0,T]$. Let $(\phi_j)_{j \in \mathbb{N}} \subset \mathcal{S}(\mathbb{R})$ be a sequence satisfying the conditions $(i)-(iii)$, needed in the definition of $B_{\infty,\infty}^s(\mathbb{S})$, and $t \in [0,T]$. Then, $Af(t)$ belongs to $C_0^{4+\alpha}(\mathbb{S})$ iff

$$\|Af(t)\|_{C^{4+\alpha}(\mathbb{S})} = \sup_{j \in \mathbb{N}} 2^{(4+\alpha)j} \left\| \sum_{k \in \mathbb{Z} \setminus \{0\}} \phi_j(k) a_k(t) e^{ikx} \right\|_{C(\mathbb{S})} < \infty.$$

Hence, instead of showing $Af(t) \in C_0^{4+\alpha}(\mathbb{S})$, we check that the sequence from the right hand side of the relation above is bounded in $\mathbb{R}$. Let $j \geq 1$

be fixed. Since $\sum_{l=-1}^{1} \phi_{l+j} = 1$ on $\operatorname{supp} \phi_j$ we get, in view of Lemma 2.2.2, that

$$\sum_k \phi_j(k) a_k(t) e^{ikx} = \sum_k \phi_j(k) a_k(t) e^{ikx} \sum_{l=-1}^{1} \phi_{l+j}(k)$$

$$= \sum_{l=-1}^{1} \int_0^t \sum_k \left[\phi_j(k) K(k, t-s)\right] \left[\phi_{l+j}(k) f_k(s) e^{ikx}\right] ds$$

$$= (2\pi)^{-1/2} \sum_{l=-1}^{1} \int_0^t \mathcal{F}^{-1}\left[\phi_j(\cdot) K(\cdot, t-s)\right] * \left[\sum_k \phi_{l+j}(k) f_k(s) e^{ik\cdot}\right](x) ds$$

$$= (2\pi)^{-1/2} \sum_{l=-1}^{1} \int_0^t \int_{\mathbb{R}_z} \mathcal{F}^{-1}\left[\phi_j(\cdot) K(\cdot, t-s)\right](x-z)$$

$$\times \sum_k \phi_{l+j}(k) f_k(s) e^{ikz} \, dz \, ds$$

$$= \sum_{l=-1}^{1} (2\pi)^{-1} \int_0^t \int_{\mathbb{R}_z} \int_{\mathbb{R}_\xi} \phi_j(\xi) K(\xi, t-s) e^{i\xi(x-z)} \, d\xi$$

$$\times \sum_k \phi_{l+j}(k) f_k(s) e^{ikz} \, dz \, ds,$$

where $*$ is the convolution symbol and the summation index $k \in \mathbb{Z}\setminus\{0\}$. We have used here the fact that the function $\phi_j(\cdot) K(\cdot, t) \in \mathcal{S}(\mathbb{R})$ for all $t \in [0, T]$, which is a consequence of the fact that ϕ_j is a rapidly decreasing function and $K(\cdot, t)$ is smooth and bounded. Hence, $\mathcal{F}^{-1}[\phi_j(\cdot) K(\cdot, t)] \in \mathcal{S}(\mathbb{R})$ and $\phi_j(\cdot) K(\cdot, t)$ can be obtained by applying the Fourier transform operator to $\mathcal{F}^{-1}[\phi_j(\cdot) K(\cdot, t)]$.

For $-1 \leq l \leq 1$, we let

$$S_{lj} := \int_0^t \int_{\mathbb{R}_z} \int_{\mathbb{R}_\xi} \phi_j(\xi) K(\xi, t-s) e^{i\xi(x-z)} \sum_k \phi_{l+j}(k) f_k(s) e^{ikz} \, d\xi \, dz \, ds.$$

We are left to estimate the $C(\mathbb{S})$–norm of S_{lj}, $-1 \leq l \leq 1$. For convenience, we write $S_{lj} := I_1 + I_2$, where

$$I_1 = \int_0^t \int_{\mathbb{R}_\xi} \int_{|z-x|\leq 2^{-j}} \phi_j(\xi) K(\xi, t-s) e^{i\xi(x-z)} \sum_k \phi_{l+j}(k) f_k(s) e^{ikz} \, dz \, d\xi \, ds,$$

and

$$I_2 = \int_0^t \int_{\mathbb{R}_\xi} \int_{|z-x|\geq 2^{-j}} \phi_j(\xi) K(\xi, t-s) e^{i\xi(x-z)} \sum_k \phi_{l+j}(k) f_k(s) e^{ikz} \, dz \, d\xi \, ds.$$

Now, since $f \in \mathbb{F}$, we obtain for each $t \in [0,T]$, $j \in \mathbb{N}$, and $l \in \{-1, 0, 1\}$ with $j + l \geq 0$ that

$$\left\|\sum_k \phi_{l+j}(k) f_k(t) e^{ikz}\right\|_{C(\mathbb{S})} \leq 2^{-(1+\alpha)(j+l)} \|f(t)\|_{h^{1+\alpha}(\mathbb{S})} \leq 2^{-(1+\alpha)(j+l)} \|f\|_\mathbb{F}.$$

We begin with the estimation of I_1:

$$|I_1| \leq 2^{-(1+\alpha)(j+l)} \|f\|_\mathbb{F} \int_0^t \int_{\mathbb{R}_\xi} \int_{|z-x|\leq 2^{-j}} |\phi_j|(\xi) K(\xi, t-s) \, dz \, d\xi \, ds$$

$$\leq 2^{-(1+\alpha)(j+l)} \|f\|_\mathbb{F} 2^{-j+1} \int_0^t \int_{\mathbb{R}_\xi} |\phi_j|(\xi) K(\xi, t-s) \, d\xi \, ds$$

$$\leq 2^{2+\alpha} \|f\|_\mathbb{F} 2^{-(2+\alpha)j} \int_{\mathbb{R}_\xi} |\phi_j|(\xi) \left(\sup_{2^{j-1}\leq |\xi|\leq 2^{j+1}} \int_0^t K(\xi, t-s) \, ds\right) d\xi$$

$$\leq 2^{2+\alpha} C \|f\|_\mathbb{F} 2^{-(5+\alpha)j} \int_{\mathbb{R}_\xi} |\phi_j|(\xi) \, d\xi$$

$$\leq 2^{2+\alpha} C \|f\|_\mathbb{F} 2^{-(5+\alpha)j} \int_{2^{j-1}\leq |\xi|\leq 2^{j+1}} c_0 \, d\xi$$

$$\leq 2^{4+\alpha} c_0 C \|f\|_\mathbb{F} 2^{-(4+\alpha)j},$$

where we used the following estimates for K. Namely, we choose $j_* \in \mathbb{N}$ with the property that $x \mapsto (\gamma/2)x^3 - \varpi x$ is positive on $[2^{j_*-1}, \infty)$. For $2^{j-1} \leq |\xi| \leq 2^{j+1}$ and $j \geq j_*$, we then have

$$\int_0^t K(\xi, t-s)\, ds = \int_0^t e^{-m(\xi)(t-s)}\, ds = \left. \frac{e^{-m(\xi)(t-s)}}{m(\xi)} \right|_0^t \leq \frac{1}{m(\xi)} \leq C 2^{-3j}. \tag{7.27}$$

Notice that for $j < j_*$, $|\xi| \leq 2^{j_*+1}$ and $t \in [0, T]$ the bound (7.27) holds also, but the constant C depends on T. The constant c_0 is chosen such that $\|\phi_j\|_0 \leq c_0$ for all $j \in \mathbb{N}$, cf. condition (iii) on $(\phi_j)_{j \geq 0}$.

We consider now the second integral I_2. Since $\phi_j \in \mathcal{S}(\mathbb{R})$ and all the other functions which appear in I_2 are smooth and have polynomial growth, we obtain after integration by parts that

$$I_2 = \int_{|z-x| \geq 2^{-j}} \int_{\mathbb{R}_\xi} \left[\left(\frac{i \partial_\xi}{x-z} \right)^2 e^{i\xi(x-z)} \right] \phi_j(\xi)$$

$$\times \left[\int_0^t K(\xi, t-s) \sum_k \phi_{l+j}(k) f_k(s) e^{ikz}\, ds \right] dz\, d\xi$$

$$= - \int_{|z-x| \geq 2^{-j}} \int_{\mathbb{R}_\xi} \frac{1}{|z-x|^2} e^{i\xi(x-z)} \sum_{m=0}^2 \binom{2}{m} \phi_j^{(2-m)}(\xi)$$

$$\times \left[\int_0^t \partial_\xi^m K(\xi, t-s) \sum_k \phi_{l+j}(k) f_k(s) e^{ikz}\, ds \right] dz\, d\xi.$$

Hence,

$$|I_2| \leq 2 \cdot 2^{-(1+\alpha)(j+l)} \|f\|_{\mathbb{F}} \int_{|z-x| \geq 2^{-j}} \frac{1}{|z-x|^2}\, dz$$

$$\times \sum_{m=0}^2 \int_{2^{j-1} \leq |\xi| \leq 2^{j+1}} \left| \phi_j^{(2-m)}(\xi) \right| \int_0^t \left| \partial_\xi^m K(\xi, t-s) \right| ds\, d\xi$$

$$\leq 2^{2j+5-(1+\alpha)j}\|f\|_{\mathbb{F}}\sum_{m=0}^{2} c_l 2^{-(2-m)j} \max_{2^{j-1}\leq |\xi|\leq 2^{j+1}} \int_0^t \left|\partial_\xi^m K(\xi,t-s)\right| ds\, d\xi.$$

Given $j \geq j_*$, $2^{j-1} \leq |\xi| \leq 2^{j+1}$ and $m \in \{0,1,2\}$, we compute that

$$\left|\partial_\xi^m K(\xi,t-s)\right| \leq C_1 |\xi|^{-m} \sum_{p=0}^{m} |m(\xi)|^p (t-s)^p K(\xi,t-s),$$

and by a suitable variable substitution, we get

$$\int_0^t \left|\partial_\xi^m K(\xi,t-s)\right| ds \leq C_1 2^m 2^{-mj} \sum_{p=0}^{m} \int_0^t |m(\xi)|^p (t-s)^p K(\xi,t-s)\, ds$$

$$\leq C_2 2^{-(m+3)j} \sum_{p=0}^{m} \int_0^\infty \tau^p e^{-\tau}\, ds \leq C_3 2^{-(m+3)j}.$$

Similarly to (7.27), we then obtain

$$\int_0^t \left|\partial_\xi^m K(\xi,t-s)\right| ds \leq C_4 2^{-(m+3)j}, \tag{7.28}$$

if $j \geq 1$ and $2^{j-1} \leq |\xi| \leq 2^{j+1}$, with a constant C_4 depending also on T. Summarising,

$$|I_2| \leq C_5 \|f\|_{\mathbb{F}} 2^{-(4+\alpha)j}.$$

For $j = 0$ we obtain similar estimates by taking into consideration that $\sum_{l=0}^{1} \phi_l = 1$ on $[|x| \leq 2]$. We have thus shown that Af is a bounded mapping, i.e. $Af \in B([0,T], C_0^{4+\alpha}(\mathbb{S}))$, and that relation (7.26) is true.

Step 2 We prove that $Af \in C_0([0,T], C_0^{4+\alpha}(\mathbb{S}))$, provided $f \in \widehat{\mathbb{F}}$. In fact, it suffices to presuppose only that $f \in C_0([0,T], C_0^{1+\alpha}(\mathbb{S}))$. Let $t \in [0,T)$ be given and choose $\delta > 0$. Set further

$$f_\delta(t) = \begin{cases} f(t+\delta) &, \ t+\delta \leq T, \\ f(T) &, \ t+\delta \geq T. \end{cases}$$

Then, $f_\delta - f \in \mathbb{F}$ for all $\delta > 0$ and $f \in \mathbb{F}$. Moreover,

$$A(f_\delta - f)(t) = \sum_{k \in \mathbb{Z}\setminus\{0\}} b_k(t) e^{ikx} \in h_0^{4+\alpha}(\mathbb{S}),$$

where

$$b_k(t) = \int_0^t \frac{\mu_-}{\mu_- + \mu_+ \tanh^2(k)} e^{-m(k)s} \left(f_k(t+\delta-s) - f_k(t-s) \right) ds.$$

We set $f_k(t) = f_k(T)$ for $t \in [T, T+\delta]$ and $k \in \mathbb{Z} \setminus \{0\}$. Furthermore,

$$Af(t+\delta) - Af(t) =$$

$$= \sum_{k \in \mathbb{Z} \setminus \{0\}} \frac{\mu_-}{\mu_- + \mu_+ \tanh^2(k)} \left(\int_0^{t+\delta} e^{-m(k)(t+\delta-s)} f_k(s)\, ds \right.$$

$$\left. - \int_0^t e^{-m(k)(t-s)} f_k(s)\, ds \right) e^{ikx}$$

$$= A(f_\delta - f)(t)$$

$$+ \sum_{k \in \mathbb{Z} \setminus \{0\}} \frac{\mu_-}{\mu_- + \mu_+ \tanh^2(k)} \int_t^{t+\delta} e^{-m(k)s} f_k(t+\delta-s)\, ds$$

The interval $[0,T]$ is compact, hence $\|f_\delta - f\|_\mathbb{F} \to 0$ for $\delta \to 0$. In view of (7.26) we then get $Af_\delta(t) - Af(t) \to 0$ in $C^{4+\alpha}(\mathbb{S})$. We are thus left to prove that

$$\sum_{k \in \mathbb{Z} \setminus \{0\}} \int_t^{t+\delta} \frac{\mu_-}{\mu_- + \mu_+ \tanh^2(k)} e^{-m(k)s} f_k(t+\delta-s)\, ds \to_{\delta \to 0} 0$$

in the $C^{4+\alpha}(\mathbb{S})$–norm. The proof of this relation is obtained analogously to that of the assertion in *Step 1* and relies strongly on the continuity of $f \in \widehat{\mathbb{F}}$ and on the fact that $f(0) = 0$. The continuity of Af from the left follows similarly.

Step 3 To complete the proof, we show that $Af \in C_0([0,T], h_0^{4+\alpha}(\mathbb{S}))$ for all $f \in \widehat{\mathbb{F}}$. Given $m \in \mathbb{N}$, it is clear from the first two steps of the proof, that $Af \in C_0([0,T], C_0^{m+3+\alpha}(\mathbb{S}))$ if $f \in C_0([0,T], C_0^{m+\alpha}(\mathbb{S}))$. Since $C_0^{2+\alpha}(\mathbb{S})$ is dense in $h_0^{1+\alpha}(\mathbb{S})$, we get, cf. [51, Corollary 0.1.3], that $C_0([0,T], C_0^{2+\alpha}(\mathbb{S}))$ is dense is $C_0([0,T], h_0^{1+\alpha}(\mathbb{S}))$.

Let $f \in \widehat{\mathbb{F}}$ be given and $(f_n) \subset C_0([0,T], C_0^{2+\alpha}(\mathbb{S}))$ a sequence with the property that $f_n \to f$ in $\widehat{\mathbb{F}}$. Then, $Af_n \to Af$ in $C_0([0,T], C_0^{4+\alpha}(\mathbb{S}))$, and additionally $Af_n(t) \in C_0^{5+\alpha}(\mathbb{S})$ for all $n \in \mathbb{N}$ and $t \in [0,T]$. Particularly, Af_n converges pointwise to Af in $C_0^{4+\alpha}(\mathbb{S})$, hence

$$Af(t) \in \overline{C_0^{5+\alpha}(\mathbb{S})}^{C^{4+\alpha}(\mathbb{S})} = h_0^{4+\alpha}(\mathbb{S}).$$

This completes the proof.

$\square$

Finally, we come to the proof of the first main result as stated in Theorem 6.2.2.

Proof (Proof of Theorem 6.2.2.) Let us recall that the function $F : \mathcal{O} \times \widehat{\mathcal{U}} \subset h_0^{4+\alpha}(\mathbb{S}) \times \widehat{\mathbb{E}} \to \widehat{\mathbb{F}}$, defined by (7.13) is, in view of Lemmas 7.1.1-7.1.2 and relation (6.6), analytic in a small neighbourhood $B_{h_0^{4+\alpha}(\mathbb{S})}(0,\epsilon) \times B_{\widehat{\mathbb{E}}}(0,r)$, for some positive constants ϵ and r. Moreover, we can choose the set $\mathcal{O}$, determined in Lemma 7.1.4, and $\epsilon > 0$ such that $\mathcal{O} = B_{h_0^{4+\alpha}(\mathbb{S})}(0,\epsilon)$.

It is well-known that the set of isomorphisms $\mathcal{L}is(\widehat{\mathbb{E}},\widehat{\mathbb{F}})$ is an open subset of $\mathcal{L}(\widehat{\mathbb{E}},\widehat{\mathbb{F}})$. Since $\partial_h F(0,0)$ is an isomorphism and F is analytic, we further assume that $\partial_h F(f,0) \in \mathcal{L}is(\widehat{\mathbb{E}},\widehat{\mathbb{F}})$ for all $f \in \mathcal{O}$. In view of $F(0,0) = 0$, we let ϵ be small enough to guarantee that the assumptions of Theorem 7.1.3 are satisfied by $\partial_h F(f,0)$ for all $f \in \mathcal{O}$ with constants k, δ, and β independent of $f \in \mathcal{O}$. Then, we find out from Theorem 7.1.3 that, given $f \in \mathcal{O}$, $F(f,h) = 0$ has a unique solution $h \in \overline{B}_{\widehat{\mathbb{E}}}(0,2\delta)$. By the implicit function theorem, this solution depends analytically on $f \in \mathcal{O}$. Consequently, $\mathcal{F}(f) := f + \mathcal{E}(f) + h$, where $h \in \overline{B}_{\widehat{\mathbb{E}}}(0,2\delta)$ is the unique solution of $F(f,h) = 0$, is a solution of (6.3), and the mapping

$$\mathcal{O} \ni f \mapsto \mathcal{F}(f) = f + \mathcal{E}(f) + h \in C([0,T], h_0^{4+\alpha}(\mathbb{S})) \cap C^1([0,T], h_0^{1+\alpha}(\mathbb{S}))$$

is, thanks to Lemma 7.1.5, analytic.

We are left to discuss the uniqueness of this solution. To this scope we solve first the following Cauchy problem. Let B be the operator defined in the proof of Lemma 7.1.5. The inhomogeneous Cauchy problem

$$\begin{cases} x' = Bx + g - Bh, & 0 \leq t \leq T, \\ x(0) = h, \end{cases} \quad (7.29)$$

possesses for each $(h, g) \in h_0^{4+\alpha}(\mathbb{S}) \times h_0^{1+\alpha}(\mathbb{S})$ a unique solution

$$x \in C([0, T], h_0^{4+\alpha}(\mathbb{S})) \cap C^1([0, T], h_0^{1+\alpha}(\mathbb{S})),$$

and

$$\max_{0 \leq t \leq T} \left(\|x(t)\|_{h_0^{4+\alpha}(\mathbb{S})} + \|\partial_t x(t)\|_{h_0^{1+\alpha}(\mathbb{S})} \right) \leq K \left(\|h\|_{h_0^{4+\alpha}(\mathbb{S})} + \|g\|_{h_0^{1+\alpha}(\mathbb{S})} \right).$$

(7.30)

For $0 < T' \leq T$ we shall write, to prevent confusion, $\widehat{\mathbb{E}}_{T'}, \widehat{\mathbb{F}}_{T'}$ and $F_{T'}$ when working with functions defined on $[0, T']$.

Let $k, \delta, \beta, \epsilon, r$ be the constants fixed in the first part of the proof. Since $\partial_h F(f, 0)^{-1} F(f, 0) \to_{f \to 0} 0$, we may choose ϵ small enough to guarantee that

$$\|\partial_h F(f, 0)^{-1} F(f, 0)\|_{\widehat{\mathbb{E}}} \leq \overline{\delta} := \frac{\delta}{K^2}$$

for all $f \in B_{h_0^{4+\alpha}(\mathbb{S})}(0, \epsilon)$. Setting $\overline{r} := r/K$, $\overline{k} := K^2 k$ we find that

$$F_{T'} \in C^\omega(B_{h_0^{4+\alpha}(\mathbb{S})}(0, \epsilon) \times B_{\widehat{\mathbb{E}}_{T'}}(0, \overline{r}), \widehat{\mathbb{F}}_{T'})$$

satisfies the assumptions of Theorem 7.1.3 for all $0 < T' \leq T$, with the constants $\overline{\beta}, \overline{\delta}, \overline{k}$ independent of T'.

Indeed, given $h \in \widehat{\mathbb{E}}_{T'}$, we shall write $\overline{h}$ for the extension of h in $\widehat{\mathbb{E}}$, which is defined by

$$\overline{h}(t) = \begin{cases} h(t) & , \quad 0 \leq t \leq T', \\ x(t - T') & , \quad T' \leq t \leq T, \end{cases}$$

and where x is the solution of (7.29) corresponding to $(h(T'), \partial_t h(T'))$. Thanks to (7.30) we find $\|\overline{h}\|_{\widehat{\mathbb{E}}} \leq K \|h\|_{\widehat{\mathbb{E}}_{T'}}$ for all $h \in \widehat{\mathbb{E}}_{T'}$. Particularly, $\|\overline{h}\|_{\widehat{\mathbb{E}}} < r$, provided $\|h\|_{\widehat{\mathbb{E}}_{T'}} < \overline{r}$. Also, given $g \in \widehat{\mathbb{F}}_{T'}$, we write $\overline{g}$ for its extension in $\widehat{\mathbb{F}}$, which is defined by

$$\overline{g}(t) = \begin{cases} g(t) & , \quad 0 \leq t \leq T', \\ g(T') & , \quad T' \leq t \leq T. \end{cases}$$

In view of relation (7.13), we see that

$$\partial_h F_{T'}(f, h_0)[h] = \partial_h F(f, \overline{h}_0)[\overline{h}] \quad \text{in} \quad [0, T']$$

for all $(f, h_0) \in B_{h_0^{4+\alpha}(S)}(0, \epsilon) \times B_{\widehat{\mathbb{E}}_{T'}}(0, \bar{r})$ and $h \in \widehat{\mathbb{E}}_{T'}$. It is also clear that we can replace $\bar{h}$ by any extension of h in $\widehat{\mathbb{E}}$, and the equality still holds. Consequently, $\partial_h F_{T'}(f, 0)$ is one-to-one. Since $\partial_h F(f, 0)$ is an isomorphism, we find for $g \in \widehat{\mathbb{F}}_{T'}$ a function $\widetilde{h} \in \widehat{\mathbb{E}}$ with $\partial_h F(f, 0)[\widetilde{h}] = \bar{g}$. Setting $h := \widetilde{h}|_{[0,T']}$, we obtain that $\partial_h F_{T'}(f, 0)[h] = g$, thus $\partial_h F_{T'}(f, 0)$ is an isomorphism, i.e.

$$\partial_h F_{T'}(f, 0) \in \mathcal{L}is(\widehat{\mathbb{E}}_{T'}, \widehat{\mathbb{F}}_{T'}) \quad \text{for all } 0 < T' \leq T \text{ and } f \in B_{h_0^{4+\alpha}(S)}(0, \epsilon).$$

In view of the relations found above, we have

$$\|\partial_h F_{T'}(f, 0)^{-1} F_{T'}(f, 0)\|_{\widehat{\mathbb{E}}_{T'}} \leq \|\partial_h F(f, 0)^{-1} F(f, 0)\|_{\widehat{\mathbb{E}}} \leq \bar{\delta},$$

$$\|\partial_h F_{T'}(f, 0)^{-1}\|_{\mathcal{L}(\widehat{\mathbb{F}}_{T'}, \widehat{\mathbb{E}}_{T'})} = \sup_{\|g\|_{\widehat{\mathbb{F}}_{T'}} = 1} \|\partial_h F_{T'}(f, 0)^{-1} g\|_{\widehat{\mathbb{E}}_{T'}}$$

$$= \sup_{\|g\|_{\widehat{\mathbb{F}}_{T'}} = 1} \|\partial_h F(f, 0)^{-1} \bar{g}\|_{\widehat{\mathbb{E}}}$$

$$\leq \sup_{\|g\|_{\widehat{\mathbb{F}}} = 1} \|\partial_h F(f, 0)^{-1} g\|_{\widehat{\mathbb{E}}} \leq \bar{\beta} := \beta,$$

and

$$\|\partial_h F_{T'}(f, h_1) - \partial_h F_{T'}(f, h_2)\|_{\mathcal{L}(\widehat{\mathbb{E}}_{T'}, \widehat{\mathbb{F}}_{T'})}$$

$$= \sup_{h \in \widehat{\mathbb{E}}_{T'} \setminus \{0\}} \frac{\|(\partial_h F_{T'}(f, h_1) - \partial_h F_{T'}(f, h_2))[h]\|_{\widehat{\mathbb{F}}_{T'}}}{\|h\|_{\widehat{\mathbb{E}}_{T'}}}$$

$$\leq K \sup_{h \in \widehat{\mathbb{E}}_{T'} \setminus \{0\}} \frac{\|(\partial_h F(f, \bar{h}_1) - \partial_h F(f, \bar{h}_2))[\bar{h}]\|_{\widehat{\mathbb{F}}}}{\|\bar{h}\|_{\widehat{\mathbb{E}}}}$$

$$\leq K \|\partial_h F(f, \bar{h}_1) - \partial_h F(f, \bar{h}_2)\|_{\mathcal{L}(\widehat{\mathbb{E}}, \widehat{\mathbb{F}})}$$

$$\leq k K^2 \|h_1 - h_2\|_{\widehat{\mathbb{E}}_{T'}}$$

for all $h_1, h_2 \in B_{\widehat{\mathbb{E}}_{T'}}(0, \bar{r})$. This is due to the fact that $\bar{h}_1, \bar{h}_2 \in B_{\widehat{\mathbb{E}}}(0, r)$ for all $h_1, h_2 \in B_{\widehat{\mathbb{E}}_{T'}}(0, \bar{r})$. Hence, Theorem 7.1.3 applies to equation $F_{T'}(f, h) = 0$ in $\widehat{\mathbb{F}}_{T'}$, and we obtain for each $f \in B_{h_0^{4+\alpha}(S)}(0, \epsilon)$ and $0 < T' \leq T$ a unique solution of $F_{T'}(f, h) = 0$ in the closed ball $\overline{B}_{\widehat{\mathbb{E}}_{T'}}(0, 2\bar{\delta})$.

We prove now that for $f \in \mathcal{O} = B_{h_0^{4+\alpha}(\mathbb{S})}(0, \epsilon)$ there exists a unique solution to the problem (6.3). Assume that we found another solution $g \in C([0,T], h_0^{4+\alpha}(\mathbb{S})) \cap C^1([0,T], h_0^{1+\alpha}(\mathbb{S}))$ of (6.3) with $g(0) = f$ and $g \neq \mathcal{F}(f)$. As we mentioned in the discussion preceding Lemma 7.1.4, $\partial_t g(0) = \mathcal{I}(f)$. Let
$$t_0 := \max\{t \in (0,T) : \mathcal{F}(f) = g \text{ on } [0,t]\} \in [0,T).$$
Since $g = \mathcal{F}(f)$ on $[0, t_0]$, we choose $T' := t_0 + \vartheta < T$, where $\vartheta > 0$ is small enough to guarantee that $\widetilde{h} = g - f - \mathcal{E}(f) \in \overline{B}_{\widehat{\mathbb{E}}_{T'}}(0, 2\overline{\delta})$. Whence, we obtain that $g = \mathcal{F}(f)$ on $[0, t_0 + \vartheta]$, in contradiction with the definition of t_0. This completes the proof. □

7.3 Nontrivial steady-states solutions of the Muskat problem

In this last section we consider the set consisting of all stationary solutions of problem (6.3). We still assume that the cell contains unit volumes of both fluids. First of all, we notice that if $f \in \mathcal{V}_0$ is a stationary solution of (6.3), then the potentials u_- and u_+ are both constant. This is due to the fact that they are both solutions of elliptic problems with homogeneous Neumann boundary conditions. We are led by the fourth equation of system (6.7) to the problem of determining the solutions of the equations

$$\gamma \frac{f''}{(1+f'^2)^{3/2}} + \varpi f = const. \quad \text{and} \quad \int_{\mathbb{S}} f \, dx = 0. \tag{7.31}$$

Recall by (6.4) that $\varpi = g(\rho_+ - \rho_-)$ is the constant which measures the density jump across the interface separating the fluids and parametrised by the function f. If the density of the fluid on the bottom of the cell is greater or equal to that of the fluid above, then (7.31) has only the trivial solution $f = 0$. Moreover, we have:

Lemma 7.3.1 *Problem* (7.31) *has at most a nontrivial solution in the set* $C^2(\mathbb{S})$ *when* $\varpi > 0$. *All the solutions of* (7.31) *are smooth.*

Proof If $\varpi = 0$, it is easy to see that also $f = 0$. Let now $\varpi < 0$ and assume that we found a solution f which is not constant zero. We find a real number $c \in \mathbb{R}$ such that

$$f - c = -\frac{\gamma}{\varpi} \frac{f''}{(1 + f'^2)^{3/2}}.$$

Since f has integral mean zero, there must be a point $x_0 \in \mathbb{S}$ such that $f(x_0) = \max f > 0$. On the one hand, if $f(x_0) > c$, then $f''(x_0) > 0$, and in view of $f'(x_0) = 0$, we deduce that x_0 is a point of minimum for f. It follows that f is constant in a neighbourhood of x_0. The set $f^{-1}(\{\max f\})$ is therefore a nonempty, closed and open set in $\mathbb{S}$. Hence f is constant, in contradiction with our assumption. On the other hand, if $f(x_0) \leq c$, then $f'' \leq 0$ in $\mathbb{S}$, which is again a contradiction. The assumption we made is then false, so that $f = 0$.

If $f \in C^2(\mathbb{S})$ is a solution of (7.31), then it follows, as in the proof of Lemma 5.3.1, that $f \in C^\infty(\mathbb{S})$. This completes the proof. □

We fix $\varpi > 0$ and look for $(\gamma, f) \in (0, \infty) \times C^2(\mathbb{S})$ with $\|f\|_{C(\mathbb{S})} < 1$ which solve (7.31). Therefore, we introduce an operator which enables us to consider both equations of (7.31) at once. Since all the solutions of (7.31) are smooth, for even functions, the problem of finding the solutions of (7.31) is equivalent to determining the solutions of the equation

$$\Upsilon(\gamma, f) = 0, \qquad (7.32)$$

in $(0, \infty) \times \mathcal{W}$, where

$$\mathcal{W} := \{f \in C^{3+\alpha}_{0,e}(\mathbb{S}) \,:\, \|f\|_{C(\mathbb{S})} < 1\}.$$

The operator $\Upsilon : \mathbb{R} \times C^{3+\alpha}_{0,e}(\mathbb{S}) \to C^\alpha_o(\mathbb{S})$ is defined to be the derivative of the left hand side of the first equation of (7.31)

$$\Upsilon(\gamma, f) := \gamma \frac{f'''}{(1 + f'^2)^{3/2}} - 3\gamma \frac{f' f''^2}{(1 + f'^2)^{5/2}} + \varpi f'$$

for $(\gamma, f) \in \mathbb{R} \times C^{3+\alpha}_{0,e}(\mathbb{S})$. Clearly, Υ depends analytically on its variables and $\Upsilon(\gamma, 0) = 0$ for all $\gamma \in \mathbb{R}$. Its Fréchet derivative $\partial_f \Upsilon(\gamma, 0)$ is a Fourier

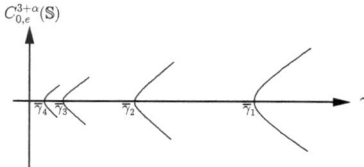

Figure 7.1: The bifurcation diagram.

multiplication operator with

$$\partial_f \Upsilon(\gamma, 0) \left[\sum_{k=1}^{\infty} a_k \cos(kx) \right] = \sum_{k=1}^{\infty} \left(\gamma k^3 - \varpi k \right) a_k \sin(kx)$$

for all $f = \sum_{k=1}^{\infty} a_k \cos(kx) \in C_{0,e}^{3+\alpha}(\mathbb{S})$. Given $l \in \mathbb{N} \setminus \{0\}$, we set

$$\overline{\gamma}_l := \frac{\varpi}{l^2}. \tag{7.33}$$

The main result of this chapter is the following global bifurcation theorem, which states that a global bifurcation branch emerges from the trivial flat solution $\{(\gamma, 0) : \gamma > 0\}$ at $(\overline{\gamma}_l, 0)$ for all $l \in \mathbb{N} \setminus \{0\}$, where $\overline{\gamma}_l$ is defined by (7.33), provided $\varpi > 0$.

Theorem 7.3.2 *Let $\varpi > 0$ and $l \geq 1$. The point $(\overline{\gamma}_l, 0)$ belongs to the closure $\mathfrak{R}$ of the set of nontrivial solutions of (7.32) in $(0, \infty) \times \mathcal{W}$. Denote by C_l the connected component of $\mathfrak{R}$ to which $(\overline{\gamma}_l, 0)$ belongs. Then C_l is not regularly bounded in $(0, \infty) \times \mathcal{W}$.*

Additionally, C_l has, in a small neighbourhood of $(\overline{\gamma}_l, 0)$, an analytic parametrisation $(\gamma_l, f_l) : (-\delta, \delta) \to (0, \infty) \times \mathcal{W}$, and

$$\gamma_l(\varepsilon) = \overline{\gamma}_l + \frac{3\varpi}{8}\varepsilon^2 + O(\varepsilon^3),$$

$$f_l(\varepsilon) = \varepsilon \cos(lx) + O(\varepsilon^2)$$

for $\varepsilon \to 0$. Moreover, any other pair $(\gamma, 0)$, with $\gamma > 0$, is not a bifurcation point.

Proof Let $l \in \mathbb{N}$, $l \geq 1$ be given. The existence of a global bifurcation branch form $(\overline{\gamma}_l, 0)$ follows using similar arguments as in the proof of Theorem 5.4.1. In virtue of Theorem 5.2.1, this bifurcation branch possesses near $(\overline{\gamma}_l, 0)$ an analytic parametrisation $(\gamma_l, f_l) : (-\delta, \delta) \to (0, \infty) \times \mathcal{W}$ with $(\gamma_l(0), f_l(0)) = (\overline{\gamma}_l, 0)$. The derivative $\gamma_l'(0)$ is zero due to the fact that $\Upsilon(\gamma, -f) = 0$ for all $(\gamma, f) \in \mathfrak{R}$. Particularly, we have that $f_l(-\varepsilon) = -f_l(\varepsilon)$ and $\gamma_l(\varepsilon) = \gamma_l(-\varepsilon)$ for all $\varepsilon \in (0, \delta)$.

We compute now the second derivative $\gamma_l''(0)$ and show that this value is positive. To this scope, for fixed $\gamma > 0$, we define $\phi : \mathbb{R}^3 \to \mathbb{R}$ by the relation

$$\phi(x, y, z) = \gamma \frac{z}{(1+x^2)^{3/2}} - 3\gamma \frac{xy^2}{(1+x^2)^{5/2}} + \varpi x \quad \text{for } (x, y, z) \in \mathbb{R}^3.$$

It holds that $\Upsilon(\gamma, f) = \phi(f', f'', f''')$ for all $f \in \mathcal{W}$. Consequently,

$$\partial_f \Upsilon(\gamma, 0)[h] = \partial_1 \phi(0) h' + \partial_2 \phi(0) h'' + \partial_3 \phi(0) h''',$$

$$\partial_f \partial_f \Upsilon(\gamma, 0)[h, h] = \partial_{11}^2 \phi(0) h'^2 + 2\partial_{12}^2 \phi(0) h' h'' + \partial_{22}^2 \phi(0) h''^2$$
$$+ 2\partial_{23}^2 \phi(0) h'' h''' + 2\partial_{13}^2 \phi(0) h' h''' + \partial_{33}^2 \phi(0) h'''^2,$$

$$\partial_f \partial_f \partial_f \Upsilon(\gamma, 0)[h, h, h] = \partial_{111}^3 \phi(0) h'^3 + 3\partial_{112}^3 \phi(0) h'^2 h'' + 3\partial_{122}^3 \phi(0) h' h''^2$$
$$+ 3\partial_{223}^3 \phi(0) h''^2 h''' + 3\partial_{113}^3 \phi(0) h'^2 h''' + 3\partial_{133}^3 \phi(0) h' h'''^2$$
$$+ 3\partial_{233}^3 \phi(0) h'' h'''^2 + 6\partial_{123}^3 \phi(0) h' h'' h''' + \partial_{222}^3 \phi(0) h''^3$$
$$+ \partial_{333}^3 \phi(0) h'''^3$$

for all $\gamma \in \mathbb{R}$ and $h \in C_{0,e}^{3+\alpha}(\mathbb{S})$. Since $\nabla \phi(0) = (\varpi, 0, \gamma)$, we obtain that $\partial_f \Upsilon(\gamma, 0)[h] = \varpi h' + \gamma h'''$, result which we already used when we determined the Fourier expansion of $\partial_f \Upsilon(\gamma, 0)[h]$. On the other side we have $\partial_{ij}^2 \phi(0) = 0$ for all $1 \leq i, j \leq 3$, so that $\partial_f \partial_f \Upsilon(\gamma, 0) = 0$. Lastly, we compute

$$\partial_{111}^3 \phi(0) = \partial_{112}^3 \phi(0) = \partial_{222}^3 \phi(0) = \partial_{223}^3 \phi(0)$$
$$= \partial_{233}^3 \phi(0) = \partial_{333}^3 \phi(0) = \partial_{123}^3 \phi(0) = \partial_{133}^3 \phi(0) = 0,$$

$$\partial_{113}^3 \phi(0) = -3\gamma, \quad \partial_{122}^3 \phi(0) = -6\gamma,$$

so that $\partial_f \partial_f \partial_f \Upsilon(\gamma, 0)[h, h, h] = -9\gamma h'^2 h''' - 18\gamma h' h''^2$ for all $\gamma > 0$ and $h \in C_0^{3+\alpha}(\mathbb{S})$.

As in the proof of Theorem 5.2.3, we write $f_l(\varepsilon) = \varepsilon \cos(lx) + \tau_l(\varepsilon)$, where $\tau_l(0) = \tau_l'(0) = 0$, and $\tau_l(\varepsilon)$ belongs to the complement of $\operatorname{Ker} \partial_f \Upsilon(\overline{\gamma}_l, 0)$ in $C^{3+\alpha}_{0,e}(\mathbb{S})$. Differentiating the relation $\Upsilon(\gamma_l(\varepsilon), \varepsilon \cos(lx) + \tau_l(\varepsilon)) = 0$ three times with respect to ε, at $\varepsilon = 0$, yields, in view of $\gamma_l'(0) = 0$ and taking into account $\partial_f \partial_f \Upsilon(\overline{\gamma}_l, 0) = 0$, that

$$\partial_f \partial_f \partial_f \Upsilon(\overline{\gamma}_l, 0)[\cos(lx)]^3 + 3\gamma_l''(0) \partial_\gamma \partial_f \Upsilon(\overline{\gamma}_l, 0)[\cos(lx)]$$
$$+ \partial_f \Upsilon(\overline{\gamma}_l, 0)[\tau'''(0)] = 0.$$

However, $\partial_f \Upsilon(\overline{\gamma}_l, 0)$ maps in the complement of $\mathbb{R} \cdot \sin(lx)$ in $C_o^\alpha(\mathbb{S})$, so that, by multiplying the relation above by $\sin(lx)$, followed by integration over the unit circle, we get

$$\gamma_l''(0) = -\frac{1}{3} \frac{\langle \partial_f \partial_f \partial_f \Upsilon(\overline{\gamma}_l, 0)[\cos(lx)]^3 \mid \sin lx \rangle}{\langle \partial_\gamma \partial_f \Upsilon(\overline{\gamma}_l, 0)[\cos(lx)] \mid \sin lx \rangle}$$

$$= -\frac{1}{3} \frac{18 l^5 \overline{\gamma}_l \int_\mathbb{S} \sin^2(ls) \cos^2(lx)\, dx - 9\overline{\gamma}_l \int_\mathbb{S} l^5 \sin^4(lx)\, dx}{l^3 \int_\mathbb{S} \sin^2(lx)\, dx}$$

$$= -3\overline{\gamma}_l l^2 \frac{\int_\mathbb{S} \sin^2(lx)(2\cos^2(lx) - \sin^2(lx))\, dx}{\int_\mathbb{S} \sin^2(lx)\, dx}$$

$$= \frac{3\overline{\gamma}_l l^2}{4} > 0.$$

In view of $\overline{\gamma}_l = \varpi/l^2$, we conclude that $\gamma_l''(0) = 3\varpi/4$.

Finally, we prove that $\mathcal{C}_1$ is relatively unbounded in $(0, \infty) \times \mathcal{W}$. The proof is based on a contradiction argument. Assume that $\mathcal{C}_1$ is relatively bounded in $(0, \infty) \times \mathcal{W}$. By repeating the arguments presented in the proof of Theorem 5.4.1, we conclude, in view of Theorem 5.4.2, that $\mathcal{C}_1$ contains some other bifurcation point $(\overline{\gamma}_k, 0)$, $k \neq 1$. Let $(\gamma_1, f_1) : [0, T] \to (0, \infty) \times \mathcal{W}$ be a continuous curve in $\mathcal{C}_1$ such that $(\gamma_1, f_1)(0) = (\overline{\gamma}_1, 0)$ and $(\gamma_1, f_1)(T) = (\overline{\gamma}_k, 0)$. Here comes into play the special structure of the equation (7.32). Given $l \in \mathbb{N}$, the continuous curve $(\gamma_l, f_l) : [0, T/l] \to (0, \infty) \times \mathcal{W}$, with

$$(\gamma_l, f_l)(\varepsilon) := (l^{-2}\gamma_1(l\varepsilon), l^{-1} f_1(l\varepsilon)(l \cdot)) \quad \text{for } \varepsilon \in [0, T/l], \tag{7.34}$$

parametrises a subset of $\mathfrak{R}$. Moreover, $(\gamma_l, f_l)(0) = (l^{-2}\overline{\gamma}_1, 0) = (\overline{\gamma}_l, 0)$ and $(\gamma_l, f_l)(T/l) = (\overline{\gamma}_{kl}, 0)$. Consequently, the union

$$\bigcup_{p=0}^\infty (\gamma_{k^p}, f_{k^p})[0, T/k^p] \subset \mathcal{C}_1,$$

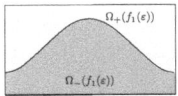 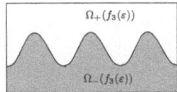 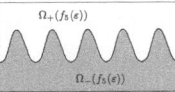

Figure 7.2: Fingering patterns on the component $\mathcal{C}_l$, $l \in \{1, 3, 5\}$.

and, taking into consideration that $\gamma_{k^p}(T/k^p) = \overline{\gamma}_{k^p} \to_{p \to \infty} 0$, we obtained a contradiction. Hence, $\mathcal{C}_1$ is relatively unbounded in $(0, \infty) \times \mathcal{W}$. Since $\mathcal{C}_l$, $l \geq 2$, is obtained from $\mathcal{C}_1$, by using the scaling (7.34), we completed the proof.

□

Notice that our analysis is simplified by the fact that $\partial_f \partial_f \Upsilon(\gamma, 0) = 0$ for all $\gamma > 0$. This is no longer the case when we compute the partial derivative $\partial_\varrho \partial_\varrho \Upsilon(\iota, 0)$ of the mapping defined by (5.10). As we shown in the proof of Theorem 5.2.3, this derivative is not the zero operator, and this is the reason why we could no determine the second derivative $\iota_l''(0)$ of the local bifurcation branch found there.

Observation 7.3.3 *For all $l \in \mathbb{N}$, $l \geq 1$, we have that $\gamma_l'(0) = 0$ and $\gamma_l''(0) > 0$. Consequently, the bifurcation is supercritical. The bifurcation diagram is pictured in Figure 7.1.*

Possible stationary fingering pattern solutions of problem (6.3) are picture in Figure 7.2. As a consequence, we obtain the following result, which states that if the fingering patterns remain bounded in $\mathcal{U}$ and the surface tension goes to infinity along a bifurcation branch, then the fingering patterns disappear.

Corollary 7.3.4 *Assume that a subset of $\mathcal{C}_l$ has a continuous parametrisation $(\gamma_l, f_l) : [0, \infty) \to (0, \infty) \times \mathcal{W}$ such that:*

(a) $\lim_{\varepsilon \to \infty} \gamma_l(\varepsilon) = \infty$,

(b) $\sup_{\varepsilon \geq 0} \|f_l(\varepsilon)\|_{C(\mathbb{S})} < 1$,

(c) $\sup_{\varepsilon \geq 0} \|f_l(\varepsilon)\|_{C^2(\mathbb{S})} < \infty$.

Then $\lim_{\varepsilon \to \infty} f_l(\varepsilon) = 0$ *in* $C^\infty(\mathbb{S})$.

Proof Given $\varepsilon \geq 0$, the pair $(\gamma_l(\varepsilon), f_l(\varepsilon))$ is a solution of (7.31), hence

$$\gamma_l(\varepsilon)\frac{(f_l(\varepsilon))''}{(1+(f_l(\varepsilon))')^{3/2}} + \varpi f_l(\varepsilon) = c(\varepsilon), \qquad (7.35)$$

where $c(\varepsilon)$ is a real constant. The assumptions we made ensure that c/γ_l is a bounded function, whence f_l is bounded in $C^k(\mathbb{S})$ for all $k \in \mathbb{N}$.

Recall, that the smooth functions build a metric space $C^\infty(\mathbb{S})$, and a sequence (f_n) converges to 0 in $C^\infty(\mathbb{S})$ iff $f_n \to 0$ in $C^k(\mathbb{S})$ for all $k \in \mathbb{N}$.

Let $\varepsilon_n \to \infty$ and $k \in \mathbb{N}$, $k \geq 2$. Assume by contradiction that $f_l(\varepsilon_n) \not\to 0$ in $C^k(\mathbb{S})$. The sequence $(f_l(\varepsilon_n))_n$ is bounded in $C^{k+1}(\mathbb{S})$, and possesses so a convergent subsequence in $C^k(\mathbb{S})$ to some $f \neq 0$. With $\varepsilon = \varepsilon_n$ in (7.35), we obtain, after dividing the relation by $\gamma_l(\varepsilon_n)$ and letting $n \to \infty$, that the curvature of $\Gamma(f)$ is constant. In view of Lemma 7.3.1, this means that $f = 0$, in contradiction with our assumption. Consequently, $f_l(\varepsilon_n) \to 0$ in $C^\infty(\mathbb{S})$ for all $\varepsilon_n \to \infty$, and we are done.

$\square$

Bibliography

[1] S. AGMON, A. DOUGLIS & L. NIRENBERG: *Estimates near the boundary for solutions of elliptic partial differential equations satisfying general boundary conditions. I*, Comm. Pure Appl. Math. **12** (1959), 623–727.

[2] D. M. AMBROSE: *Well-posedness of two-phase Hele-Shaw flow without surface tension*, European J. Appl. Math., **15** (2004), 597–607.

[3] N. D. ALIKAKOS, P. W. BATES & XINFU CHEN: *Convergence of the Cahn-Hilliard equation to the Hele-Shaw model*, Arch. Rational Mech. Anal. **128** (1994), 165–205.

[4] H. AMANN: *Linear and Quasilinear Parabolic Problems, Volume I*, Birkhäuser, Basel, 1995.

[5] H. AMANN & J. ESCHER: *Analysis III*, Birkhäuser, Basel, 2009.

[6] W. ARENDT & S. BU: *Operator-valued Fourier multipliers on periodic Besov spaces and applications*, Proc. Edinburgh Math. Soc. **47** (2004), 15–33.

[7] R. ARIS: *Vectors, Tensors, and the Basic Equations of Fluid Mechanics*, Prentice-Hall, London, 1962.

[8] G. ASTARITA & G. MARRUCCI: *Principles of Non-Newtonian Fluid Mechanics*, McGraw-Hill, London, 1974.

[9] J. BEAR & Y. BACHMAT: *Introduction to Modeling of Transport Phenomena in Porous Media*, Kluwer Academic Publisher, Boston, 1990.

[10] D. BOTHE & J. PRÜSS: L_p- *theory for a class of non-Newtonian fluids*, SIAM J. Math. Anal., **39**(2) (2007), 379-421.

[11] D. E. BOURNE & P. C. KENDALL: *Vector Analysis and Cartesian Tensors*, Chapman & Hall, London, 1992.

[12] BUFFONI, B., TOLAND, J.: *Analytic Theory of Global Bifurcation: An Introduction*, Princeton, New Jersey, 2003.

[13] L. CARRILLO, F. X. MAGDALENO, J. CASADEMUNT & J. ORTÍN: *Experiments in a rotating Hele–Shaw cell*, Phys. Rev. E, **54**(6) (1996), 6260–6267.

[14] L. CARRILLO, J. SORIANO & J. ORTÍN: *Interfacial instabilities of a fluid annulus in a rotating Hele–Shaw cell*, Phys. Fluids, **12**(7) (2000), 1685–1698.

[15] C.-Y. CHEN & S.-W. WANG: Interfacial instabilities of miscible fluids in a rotating Hele–Shaw cell, *Fluid Dynam. Res.*, **30** (2002), 315–330.

[16] XINFU CHEN: *The Hele-Shaw problem and area-preserving curve-shortening motions*, Arch. Rational Mech. Anal., **123** (1993), 117–151.

[17] M. G. CRANDALL & P. H. RABINOWITZ : *Bifurcation from simple eigenvalues*, J. Funct. Anal., **8** (1971), 321–340.

[18] S. CUI & J. ESCHER: *Bifurcation analysis of an elliptic free boundary problem modelling the growth of avascular tumors*, SIAM J. Math. Anal. **39**(1) (2007), 210–235.

[19] G. DA PRATO & P. GRISVARD: *Equations d'évolution abstraites non-linéaires de type parabolique*, Ann. Mat. Pura Appl., **120** (1979), 329–326.

[20] K. DEIMLING: *Nonlinear Funktional Analysis*, Springer, Berlin, 1985.

[21] M. DOBROWOLSKI: *Angewandte Funktionalanalysis. Funktionalanalysis, Sobolev-Räume und elliptische Differentialgleichungen*, Springer, Berlin, 2006.

[22] M. EHRNSTRÖM, J. ESCHER & B-V. MATIOC: *Well-posedness, instabilities, and bifurcation results for the flow in a rotating Hele-Shaw cell*, submitted.

[23] J. ESCHER, A.-V. MATIOC & B.-V. MATIOC, : *Classical solutions and stability results for Stokesian Hele-Shaw flows*, Ann. Scuola Norm. Sup. Pisa Cl. Sci., to appear.

[24] J. ESCHER & B.-V. MATIOC: *Stability of the equilibria for periodic Stokesian Hele-Shaw flows*, J. Evol. Equ., **8**(3) (2008), 513–522.

[25] J. ESCHER & B.-V. MATIOC: *Existence and stability results for periodic Stokesian Hele-Shaw flows*, European J. Appl. Math., **40**(5) (2008), 1992–2006.

[26] J. ESCHER & B.-V. MATIOC: *On periodic Stokesian Hele-Shaw flows with surface tension*, SIAM J. Math. Anal., **19**(6) (2008), 717–734.

[27] J. ESCHER & B.-V. MATIOC: *A moving boundary problem for periodic Stokesian Hele-Shaw flows*, Interfaces Free Bound., **11** (2009), 119–137.

[28] J. ESCHER & B.-V. MATIOC: *Multidimensional Hele-Shaw flows modeling Stokesian fluids*, Math. Methods Appl. Sci., **32** (2009), 577–593.

[29] J. ESCHER & G. PROKERT: *Stability of the equilibria for spatially periodic flows in porous media*, Nonlinear Anal., **45** (2001), 1061–1080.

[30] J. ESCHER & G. SIMONETT : *Classical solutions for Hele-Shaw models with surface tension*, Adv. Differential Equations, **2** (1997), 619–642–1047.

[31] J. ESCHER & G. SIMONETT : *Classical solutions of multidimensional Hele–Shaw models*, SIAM J. Math. Anal., **28**(5) (1997), 1028–1047.

[32] J. ESCHER – G. SIMONETT: *A center manifold analysis for the Mullins-Sekerka model*, J. Differential Equations, **143** (1998), 267–292.

[33] P. FAST, L. KONIC, M. S. SHELLEY & P. PALFFY-MUHORAY: *Pattern formation in non-Newtonian Hele-Shaw flow*, Phys. Fluids, **13** (2001), 1191–1212.

[34] R. FOLCH, E. ALVAREZ-LACALLE, J. ORTÍN, J. & J. CASADEMUNT: *Spontaneous pinch-off in rotating Hele-Shaw flows*, Preprint.

[35] A. FRIEDMAN & Y. TAO: *Nonlinear stability of the Muskat problem with capillary pressure at the free boundary*, Nonlinear Anal. **53** (2003), 45–80.

[36] P. R. GARABEDIAN: *On steady-state bubbles generated by Taylor instability*, Proc. R. Soc. Lond. A, **241** (1957), 423–431.

[37] D. GILBARG & T. S. TRUDINGER: *Elliptic Partial Differential Equations of Second Order*, Springer–Verlag, New York, 1998.

[38] H. S. HELE-SHAW: *Investigation of the nature of surface resistance of water and of stream line motion under certain experimental conditions*, Trans. Inst. Nav. Archit., (1898).

[39] H. HEUSER: *Funktionalanalysis*, Teubner, Stuttgart, 1986.

[40] R. Y. HONG, Z. Q. REN, Y. P. HAN, H. Z. LI, Y. ZHENG & J. DING: *Rheological properties ofwater-based Fe3O4 ferrofluids*, Chemical Engineering Science, **62** (2007), 5912-5924.

[41] J. K. HUNTER & B. NACHTERGAELE: *Applied Analysis*, World Scientific, Singapore, 2001.

[42] D. P. JACKSON & J. A. MIRANDA : *Controlling fingering instabilities in rotating ferrofluids*, Phys. Rev. E, **67** (2003), 017301-1–017301-4.

[43] L. JIANG & Y. CHEN: *Weak formulation of multidimemsional Muskat problem*, K. H. Hoffmann, J. Sprekels (Eds.), Free Boundary Problems: Theory and Applications, (1990), 509-513.

[44] R. KAISER & E. R. ROSENSWEIG: *Study of Ferromagnetic Liquid*, National Aeronautics and Space Administration, Washington, D.C., 1969.

[45] T. KATO: *Perturbation Theory for Linear Operators*, Springer–Verlag, Berlin Heidelberg, 1995.

[46] H. KIELHÖFER: *Bifurcation Theory: An Introduction with Applications to PDEs*, Springer–Verlag, New York, 2004.

[47] L. KONIC, M. S. SHELLEY & P. PALFFY-MUHORAY: *Models of non-Newtonian Hele-Shaw flow*, Phys. Rev. E, **54**(5) (1996), R4536–R4539.

[48] O. A. LADYZHENSKAYA: *The Mathematical Theory of Viscous Incompressible Flow*, Gordon and Beach, New York, 1969.

[49] O. A. LADYZHENSKAYA & N. N. URALTSEVA: *Linear and Quasilinear Elliptic Equations*, Academic Press, New York, 1968.

[50] E. S. G. LEANDRO, R. M. OLIVEIRA, J. A. MIRANDA: *Geometric approach to stationary shapes in rotating Hele–Shaw flows*, Phys. D, **237** (3) (2008), 652–664.

[51] A. LUNARDI: *Analytic Semigroups and Optimal Regularity in Parabolic Problems*, Birkhäuser, Basel, 1995.

[52] J. A. MIRANDA: *Rotating Hele-Shaw flows with ferrofluids*, Phys. Rev. E, **62** (2) (2000), 2985–2988.

[53] J. A. MIRANDA & E. ALVAREZ-LACALLE: *Viscosity contrast effects on fingering formation in rotating Hele-Shaw flows*, Phys. Rev. E, **72** (2005), 026306.

[54] J. A. MIRANDA & M. WINDOM: *Stability analysis of polarized domains*, Phys. Rev. E, **55** (3) (1997), 3758–3761.

[55] M. MUSKAT: *Two fluid systems in porous media. The encroachment of water into an oil sand*, Physics, **5** (1934), 250–264.

[56] R. M. OLIVEIRA & J. A. MIRANDA: *Ferrofluid patterns in a radial magnetic field: Linear stability, nonlinear dynamics, and exact solutions*, Phys. Rev. E, **77** (2008), 016304.

[57] E. ROSENSWEIG: *Ferrohydrodynamics* Dover, New York, 1997.

[58] P. G. SAFFMAN & G. I. TAYLOR: *The penetration of a fluid into a porous medium or Hele–Shaw cell containing a more viscous fluid*, Proc. R. Soc. A, **245** (1958), 312–329.

[59] H.–J. SCHMEISSER & H. TRIEBEL: *Topics in Fourier Analysis and Function Spaces*, John Wiley and Sons Limited, New York, 1987.

[60] M. SIEGEL, R. E. CAFLISCH & S. HOWISON: *Global Existence, Singular Solutions, and Ill-Posedness for the Muskat Problem*, Comm. Pure Appl. Math., **57** (2004), 1374–1411.

[61] E. SINESTRARI: *On the abstract Cauchy problem of parabolic type in spaces of continuous functions*, J. Math. Anal. Appl., **107** (1985), 16–66.

[62] S. TANVEER: *Analytic theory for the selection of a symmetric Saffman-Taylor finger in a Hele-Shaw cell*, Phys. Fluids, **30** (1997), 56–65.

[63] E. T. TOORMAN: *Modelling the thixotropic behaviour of dense cohesive sediment suspensions*, Rheol. Acta, **36** (1987), 1589–1605.

[64] C.-Y. WEN, C.-Y. CHEN & D.-C. KUAN: *Experimental studies of labyrinthine instabilities of miscible ferrofluids in a Hele-Shaw cell*, Phys. Fluids, **19** (2007), 084101.

[65] X. XIE & S. TANVEER: *Rigorous results in steady finger selection in viscous fingering*, Arch. Rational Mech. Anal., **166** (2003), 219–286.

[66] F. YI: *Local classical solution of Muskat free boundary problem*, J. Partial Diff. Eqs., **9** (1996), 84–96.

[67] F. YI: *Global classical solution of Muskat free boundary problem*, J. Math. Anal. Appl., **288** (2003), 442–461.

[68] F. ZHOU, J. ESCHER & S. CUI: *Bifurcation for a free boundary problem with surface tension modeling the growth of multi-layer tumors*, J. Math. Anal. Appl., **337** (1) (2008), 443–457.

Die VDM Verlagsservicegesellschaft sucht für wissenschaftliche Verlage abgeschlossene und herausragende

Dissertationen, Habilitationen, Diplomarbeiten, Master Theses, Magisterarbeiten usw.

für die kostenlose Publikation als Fachbuch.

Sie verfügen über eine Arbeit, die hohen inhaltlichen und formalen Ansprüchen genügt, und haben Interesse an einer honorarvergüteten Publikation?

Dann senden Sie bitte erste Informationen über sich und Ihre Arbeit per Email an *info@vdm-vsg.de*.

Sie erhalten kurzfristig unser Feedback!

VDM Verlagsservicegesellschaft mbH
Dudweiler Landstr. 99
D - 66123 Saarbrücken

Telefon +49 681 3720 174
Fax   +49 681 3720 1749

www.vdm-vsg.de

Die VDM Verlagsservicegesellschaft mbH vertritt

Printed by Books on Demand GmbH, Norderstedt / Germany